Isabel Kraut

Separating the Aerosol Effect in Case of a „Medicane"

Wissenschaftliche Berichte des Instituts für Meteorologie und Klimaforschung des Karlsruher Instituts für Technologie (KIT)
Band 68

Herausgeber: Prof. Dr. Ch. Kottmeier

Institut für Meteorologie und Klimaforschung
am Karlsruher Institut für Technologie (KIT)
Kaiserstr. 12, 76128 Karlsruhe

Eine Übersicht aller bisher in dieser Schriftenreihe erschienenen Bände finden Sie am Ende des Buches.

Separating the Aerosol Effect in Case of a „Medicane“

by
Isabel Kraut

KIT Scientific Publishing

Dissertation, Karlsruher Institut für Technologie (KIT)
Fakultät für Physik, 2015
Referenten: Prof. Dr. Christoph Kottmeier, Prof. Dr. Corinna Hoose

Impressum

KIT Scientific Publishing

Karlsruher Institut für Technologie (KIT)
KIT Scientific Publishing
Straße am Forum 2
D-76131 Karlsruhe

www.ksp.kit.edu

Print on Demand 2015

ISSN 0179-5619
ISBN 978-3-7315-0405-4
DOI: 10.5445/KSP/1000047813

Separating the Aerosol Effect in Case of a "Medicane"

Zur Erlangung des akademischen Grades eines

DOKTORS DER NATURWISSENSCHAFTEN

von der Fakultät für Physik des

Karlsruher Instituts für Technologie (KIT)

genehmigte

DISSERTATION

von

Dipl.-Met. Isabel Kraut

aus Bad Friedrichshall

Tag der mündlichen Prüfung:	17. April 2015
Referent:	Prof. Dr. Christoph Kottmeier
Korreferent:	Prof. Dr. Corinna Hoose

Contents

Abstract

This thesis addresses the interactions of aerosols, clouds and dynamics in case of a so-called Medicane. This type of cyclone occurs over the Mediterranean Sea, showing similarities to Hurricanes over the Atlantic and Pacific Ocean. Due to the high wind speed of the Medicane, a large amount of sea salt particles is emitted over the sea. Thus, the Medicane creates its own sea salt field, which is a particular feature and can influence the development of the Medicane, its associated clouds, and precipitation.

A distinct Medicane that occurred in early November 2011 is simulated with the modeling system COSMO-ART. To investigate the impact of sea salt particles on this Medicane, two different aerosol scenarios are defined. The influence of the aerosol scenarios is further investigated with the help of two distinct ensemble simulations. In both ensembles the initial conditions are varied in different ways. The first ensemble uses a preexisting ensemble as driver for the simulations. The second ensemble is generated by adding a random perturbation on the time derivative of temperature at the first time step. In particular, these ensembles are designed to extract the aerosol effect on clouds and precipitation from effects due to the non-linearity of dynamics. This non-linearity can cause significant differences in a system's response to small changes in initial conditions. Processes contributing to this non-linearity are for example convection and cloud micro-physics. Non-linearity occurs in nature as well as in the model's simplified equation system.

For the analysis, a score was applied that does not take the difference between two model results, which is a common practice. This score is called Displacement and Amplitude Score (DAS). It uses an optical-flow algorithm to compare two precipitation fields and to quantify how well the features are matched in terms of location and intensity. DAS was originally developed to validate forecasts by a comparison with satellite data. In contrast to previous studies, this is the first study that applies DAS on different model results. The combination of DAS and a Factorial Method is a new methodology designed as a part of this thesis in order to separate the effects of aerosol on precipitation.

The application of the new methodology on the ensemble simulations of the Medicane shows, that sea salt aerosols produced spatial shifts and differences in intensity of surface precipitation. Most of the shifts assigned to aerosol interactions were between 6 and 20 km within 3 h. Changes in the precipitation amount up to 12 mm were common in the same time range. Impacts of aerosol on precipitation were determined up to a lead time of 1 day for amplitude effects and up to 1.5 days for displacements. Beyond 1.5 days of simulation, distinguishing the impact of aerosols from the impacts of other factors was not possible.

1. Introduction

The effect of liquid and solid particles in the atmosphere (aerosol) on cloud formation and precipitation is one of the least understood processes in atmospheric sciences. This holds for climate projections as the fifth Assessment Report of the Intergovernmental Panel on Climate Change (IPCC, Stocker et al., 2013) shows, but also for precipitation forecast on time scales of hours to days (e.g. Noppel et al., 2010). Although big effort has been spent in observational (e.g. Loeb and Schuster, 2008) as well as in modeling studies (e.g. Lohmann and Feichter, 2005; Ekman et al., 2006) in the past years, the interactions of aerosols, clouds, and precipitation are quantitatively not well understood (Levin and Cotton, 2008; Stevens and Feingold, 2009; Stevens and Boucher, 2012; Tao et al., 2012).

Aerosols influence the energy fluxes due to scattering and absorption of radiation. Additionally, aerosols affect cloud formation when it acts as cloud condensation nuclei or ice nuclei. This has an impact on precipitation and thus on the water cycle. Cloud micro-physical processes can compensate each other, which makes it difficult to quantify the relative importance of single processes for precipitation. Thus, Stevens and Feingold (2009) proposed to study the effects of aerosols on clouds and precipitation for different cloud regimes rather than performing isolated cause-effect studies. A difficulty in those studies lies in the model deficiencies. One example of a model deficiency is the need for parameterizations due to the choice of grid mesh size. Typical values are a few kilometers for numerical weather prediction up to a few hundreds kilometers on global climate studies. Thus, processes like convection and precipitation cannot be resolved and must be parameterized. This causes large uncertainties, as closure assumptions and the choice of parameters to solve those equations have to be made (Grell and Freitas, 2014).

This thesis addresses the feedback between aerosols and a special weather phenomenon over the Mediterranean Sea called a Medicane. Thus, this study has its focus on time scales of hours to days. Medicanes (from "Mediterranean hurricanes") are cyclones with characteristics of hurricanes: they have a cloud-free core surrounded by an eye-wall with a roughly axis-symmetric cloud pattern. The occurrence of Medicanes is estimated to be one in two years (Tous and Romero, 2013) up to two in one year (Cavicchia and von Storch, 2012). Medicanes are associated with strong winds and heavy precipitation, causing flooding and high surf breaks during landfall. The risk includes loss of live and property damage. The high wind speed can cause large emissions of sea salt particles. In this way, the Medicane creates its own sea salt field and keeps it alive as long as the wind speeds are high. The sea salt particles can then affect the Medicane's development and its associated clouds. Thus, the system of Medicane and sea salt particles forms a feedback circuit. For this specific weather phenomenon, interactions and the coupling of aerosol, precipitation, and wind flow are expected to be particularly strong.

Previous studies of the feedback between aerosol and precipitation on a regional scale could not consistently distinguish the impact of aerosol from other factors. Methods applied in those studies include determining difference in the precipitation fields of two scenarios with different (often constant) CCN numbers or comparing cumulative precipitation or mean values over distinct areas. In the latter method, information on the temporal and spatial distributions is lost; the former method may unintentionally represent the non-linear behavior of dynamics rather than aerosols. Non-linear systems are typically sensitive to small changes in initial conditions, which can cause the system to change its temporal evolution significantly. Processes like convection and cloud micro-physics are parts of the non-linear dynamics in the atmosphere and interfere with aerosol. This accentuates the need for improved methods to study the feedback of aerosol and precipitation. The first objective of this thesis is as follows:

1. *How can aerosol effects on clouds be determined? Can a better method be developed for separating the effects of aerosol from those of the non-linear behavior of the dynamics?*

A new method is applied on a Medicane, which occurred in November 2011. The Medicane's strong winds produced high emissions of sea salt particles. Thus, a Medicane produces and feeds its own sea salt field, which can affect its cloud structure and associated precipitation. I am not aware of any other studies on the impacts of aerosols on Medicanes. This case study does not allow to draw generalized conclusions on the impact of aerosol on precipitation. But the choice of a phenomenon where a strong coupling is expected shows evidence of maximum effects on time scales relevant for weather predictions. The research questions are as follows:

2. *What are the effects of sea salt particles on a Medicane, particularly its precipitation?*
   - *Do particles lead to spatial and temporal shifts in precipitation because the onset of precipitation formation is delayed or accelerated? Do sea salt particles change the precipitation amount?*
   - *How large are spatial shifts in precipitation? How fast does the impact of sea salt particles on precipitation propagate (i.e. how large is this signal speed)? Is the signal coupled to the horizontal wind velocity, the medicane's transition velocity, or neither?*
   - *On what time scales can aerosol effects be observed in the simulations? Is aerosol observability dependent on the lifetime of single clouds (minutes to a few hours) or on the lifetime of the entire system (a few days)?*

The dissertation is structured as follows. It provides the state of the art of aerosol–cloud interactions, an overview of the occurrence of aerosol over the Mediterranean Sea, and a description of the phenomenon Medicane in chapter 2. Chapter 3 introduces the modeling system used for the simulations. Chapter 4 gives the synoptical analysis of Medicane Rolf, the simulations setup and validation of the input data for forcing COSMO-ART at the lateral boundaries. The chapter ends with a comparison of the modeling results with observational data and a presentation of the simulated characteristics to verify the model's ability to properly simulate the Medicane. In chapter 5, the methods to create the ensembles and an improved methodology to analyze the data are presented. The application of the method is demonstrated in chapter 6.

2. On Aerosol, Clouds, and Mediterranean Characteristics

This chapter summarizes recent research on aerosol–cloud interactions with a focus on modeling studies. An overview of aerosol types and typical mass and number concentrations over the Mediterranean Sea is given. The last section within this chapter describes findings on Medicanes.

2.1. Aerosol – Cloud Interactions

Aerosols alter the energy budget and influence cloud formation, hence impact precipitation and water cycles. This happens because of two major effects. One is the direct scattering and absorption of radiation by aerosols. The scattering of solar radiation by aerosols result in cooling of the upper troposphere, whereas absorption results in warming at upper levels but cooling near the surface. The efficiency of scattering and absorption strongly depends on the composition of aerosol. For example soot particles are effective at absorbing radiation, and aerosols consisting of sulfate are more effective at scattering. The other major effect impacts the energy budget and water cycles in an indirect way, when aerosols act as cloud condensation nuclei or ice nuclei. The following paragraphs describe that impact in more detail.

2.1.1. Activation of Aerosols to Cloud Condensation Nuclei

Aerosol particles consist of hygroscopic compounds that can condense water vapor onto their surfaces, which is why aerosols are called condensation nuclei (CN). Cloud condensation nuclei (CCN) is the portion of CN that is capable of initiating cloud droplet formation. This process is described by Köhler theory (Seinfeld and Pandis, 2012), in which the equilibrium saturation over a solution droplet surface is a function of the droplet radius. Köhler theory includes two effects that consider the surface curvature (Kelvin effect, Seinfeld and Pandis, 2012) and the chemical composition of the droplet (Raoult's law, Seinfeld and Pandis, 2012). The Kelvin effect states that the water vapor pressure is higher over a curved surface than over a flat surface. Raoult's law states that a solution droplet reduces the water vapor pressure compared with a droplet that consists of pure water. The Köhler curve (derived from Köhler theory) peaks at the so-called critical saturation. This critical saturation depends on the CN's radius, solute amount and composition, and temperature. If the ambient saturation exceeds the critical saturation, then the CN can be activated as a cloud droplet. Otherwise, the particle remains un-activated. This shows that whether CN can act as CCN depends on the properties of the particle itself and on the ambient atmospheric conditions.

Simulations that aim to study the impact of aerosols on warm clouds show that in a clean environment with a low number of CCN, fewer droplets with larger diameters are produced (e.g., Tao et al., 2007). In more polluted environments, more droplets with smaller diameters and narrower size spectra are simulated (e.g., in the simulation of convective storms by van den Heever et al., 2006). More droplets scatter solar radiation back to the atmosphere and brighten the cloud ("cloud albedo effect" Twomey, 1977). The simulations agree reasonably with observations (e.g., Squires, 1958; Twomey et al., 1984; Albrecht, 1989; Rosenfeld, 1999). Teller and Levin (2006) simulated an increase in the cloud lifetime for an increase in the number of CCN due to the suppression of precipitation in case of subtropical clouds. This findings agrees with the observations of the "cloud lifetime effect" (Albrecht, 1989; Ackerman et al., 2000). However, even if warm rain is suppressed, the lifetime of clouds does not inevitably increase, as shown by the simulations of shallow cumulus by Jiang et al. (2006). These authors proposed that the suppression of warm rain competes with enhanced evaporation which shortens a cloud's lifetime. Thus, the aerosol–cloud interaction is complex, even in the warm phase. In nature as well as in modeling studies, the effects cannot be separated. The effects may mutually reinforce or counteract each other, which results in an unexpected system response (Stevens and Feingold, 2009; Seifert et al., 2012).

2.1.2. Effect of Cloud Condensation Nuclei on Rain Formation

Once initiated, cloud droplets can grow until their sedimentation velocity is large enough to leave the cloud again. If no ice crystals are involved in the rain formation, the process is called warm-rain formation. The interaction of aerosol and warm-rain formation may be strong. Cloud droplets compete for the available water vapor within the cloud, which is why more cloud droplets inevitably lead to smaller droplets. Because smaller droplets have lower sedimentation velocities and a smaller collision efficiency, growth of droplets due to collision coalescence is not as efficient as for larger droplets. Increasing the number of cloud droplets narrows its droplet size spectrum, which further reduces the collision coalescence efficiency, because significant relative motions are needed, but a narrow spectrum results in similar fall speeds. Condensation is another process which leads to the growth of cloud droplets to rain drop size, but its contribution is small in comparison to that of collision coalescence. When cloud droplets reach rain drop size, they grow mainly by the collection of cloud droplets, which also depends on the cloud droplet size. All effects described in this paragraph result in a suppression of warm-rain formation for higher CN concentrations.

Various modeling studies were performed to account for changes in surface precipitation due to altered numbers of CCN. The responses of the models to an enhanced number of CCN have discrepancies: some show an increase in precipitation (e.g., Phillips et al., 2002; Khain et al., 2004, 2005; Teller and Levin, 2006) whereas others show a decrease (e.g., Wang, 2005; Khain et al., 2005; Lee et al., 2009). Besides the atmospheric conditions, the cloud type is also responsible for the impacts of aerosols on precipitation (Seifert and Beheng, 2006; Khain et al., 2008; van den Heever et al., 2011). For single isolated cumulus clouds, enhanced CCN numbers usually result in a decrease in precipitation, whereas more complex cloud systems and deep convective clouds produce different responses. There are several mechanisms that can explain the enhanced rainfall for higher numbers of

CCN (Tao et al., 2012). The "latent heat–dynamic effect" indicates that more latent heat can be released while CCN suppress rainfall because more liquid water is retained in the cloud. The result is stronger updrafts and additional water transport to high levels of the cloud that may be frozen. This effect is particularly important in deep convective clouds (e.g., Khain et al., 2005; Van Den Heever and Cotton, 2007). Another responsible mechanism is the strong evaporative cooling due to the enhanced number of small cloud droplets. Strong evaporative cooling can strengthen the near-surface cold pool, which enhances convergence and convection. This "cold pool effect" is mainly important over the ocean in the development of convective systems (Lee et al., 2009). Another mechanism indicates that for higher CCN, a greater increase in the total water content can occur within a cloud (Wang, 2005). This process could also lead to more vigorous convection. In this case, rain is mainly produced from the ice phase, which is why this mechanism is called the "cold micro-physics effect". Of course, these processes affect each other and therefore cannot be separated. Because their feedbacks have different signs, they can compensate for each other and can result in a weaker effect than expected.

To consider this "buffering" of single processes, Stevens and Feingold (2009) propose to study the effects of aerosols on clouds and precipitation for individual cloud regimes rather than concentrating on isolated effects. Lee et al. (2009), van den Heever et al. (2011), and Rieger et al. (2014) also performed such modeling studies at the regional scale.

Giant CCN, in the range of micrometers, can accelerate rain formation. Because of their large size, these CCN can be directly activated as rain drops to initiate rain formation. These giant CCN also compete for available water vapor and lower the maximum saturation of the ambient air; thus smaller particles have difficulty becoming activated. Thus, the number of cloud droplets is smaller compared to a situation with enhanced CCN in the smaller size range. Droplets can grow more efficiently at a lower number concentration by collision–coalescence and can reach rain drop size earlier. This process counteracts the suppression of warm-rain formation described earlier. Natural giant CCN are, for example, large sea salt particles or mineral dust with hygroscopic coating. The effect of giant CCN on precipitation has been taken advantage of by adding giant CCN to the air to attempt to control rain formation (e.g., Mather et al., 1997).

2.1.3. Effect of Ice Nuclei on Cold-Cloud Processes

Aerosol particles can also act as ice nuclei (IN). With the assistance of IN, heterogeneous nucleation can occur (a process that leads to cloud glaciation). These processes are not well understood compared with the activation of CN as cloud droplets. Mineral dust has a high ice nucleating efficiency (Isono et al., 1959). Biological aerosols such as pollen, fungi, spores, and bacteria are very efficient IN (Hoose et al., 2010; Hummel et al., 2014), but their concentrations are usually quite low compared with mineral dust. Soot from natural or anthropogenic burning is an IN with a weaker efficiency.

There are four processes through which IN can initiate ice formation (Tao et al., 2012). Mainly steered by supersaturation, water can be deposited onto IN as ice (deposition nucleation), or it can be deposited as liquid and freeze afterwards (condensation-freezing nucleation). Processes that are mainly steered by temperature are immersion freezing, when an IN enters a (liquid) cloud droplet and initiates freezing at the appropriate temperature, and contact freezing, which initiates freezing of a super-cooled droplet by collision with an IN. On the global scale, Hoose et al. (2010) examined immersion freezing of mineral dust as the dominant ice formation process.

Once ice crystals are formed, they first grow by the deposition of water vapor, which is the evaporation of cloud droplets (Wegener–Bergeron–Findeisen process), or by cooling the air through lifting. When ice crystals are large, they also grow by riming, in which cloud water is accreted. All of these processes strongly depend on the size and number of cloud particles and thus on the number of CN. Jiang et al. (2008) examined the influence of aerosols on ice clouds according to the moisture content. In a wet environment, aerosols have little impact on the crystal size and precipitation. In a dry environment, aerosols reduce the ice crystal size and reduce precipitation. If more cloud droplets reach the zero degree Celsius isotherm (because the enhanced aerosol number concentration reduces the size of cloud droplets and suppresses warm-rain formation), then the growth of large hail and the cold-rain process are enhanced. This effect is called "aerosol invigoration effect" (Andreae et al., 2004; Rosenfeld et al., 2008). In a polluted environment, a cloud can grow taller than in a clean environment. This effect depends, similar to the other aerosol–cloud effects, on the cloud height and the location of the cloud within the atmosphere.

A few modeling studies have been performed to study the impact of IN on precipitation. van den Heever et al. (2006) found that higher numbers of IN produce ice at warmer temperatures and generate deeper anvils. At the beginning of cloud formation, more precipitation is produced with higher numbers of IN; later, the cumulative precipitation is greatest for the case with the lowest aerosol concentrations. Ekman et al. (2007) found that higher numbers of IN lead to enhanced updrafts because of the latent heat release from the growth of ice crystals by diffusion. The authors also suggest that this process leads to an enlarged anvil coverage and enhanced precipitation. This finding strongly agrees with the simulations by Fan et al. (2010) and with the observations by Heymsfield et al. (2005).

2.1.4. Impacts of Aerosols on Tropical Cyclones

Ebert et al. (2003) show that it is difficult for numerical weather prediction models to precisely predict the time and location of heavy precipitation events (larger than 20 mm per day). To my knowledge, the impact of aerosols on Medicanes has not been studied until now. However, the scientific interest in the impact of aerosols on tropical cyclones has increased over the last two decades. Dunion and Velden (2004) and Evan et al. (2006) studied the relationship of the Saharan air layer on tropical cyclone activity using satellite observations. These authors found that cyclones weaken if the Saharan air layer is present due to the intrusion of dry and stable air, the enhancement of the local vertical wind and the enhancement of the trade wind inversion, which all stabilize the environment. Zhang et al. (2009) suggest that convection in the eyewall decreases because there are fewer aerosols, while convection in the outer bands increases. The result is upper-level divergence from the outer bands and suppressed eyewall convection (cyclone decay). Rosenfeld et al. (2012) show that an enhanced number of CCN slows rain formation as more water reaches the freezing level. More latent heat is released to invigorate convection, with greater electrification and lightning. The higher ice-based precipitation cools the air that converges into the center of the storm; in addition to the arguments by Zhang et al. (2009), this process further weakens the cyclone. The authors also show that due to the strong winds, the emission of large sea salt particles forces the cloud to rain out early on and offset the delay in the rain formation. Hazra et al. (2013) modeled a weak cyclone for a high aerosol load. The authors found no impact of aerosols on the track of the cyclone. Wang et al. (2014) found that tropical cyclone development was delayed, the cyclone intensity was weak, the cyclone dissipated early, the rainbands expanded and the surface precipitation increased for a case with an enhanced number of aerosols. A climatic study by Dunstone et al. (2013) suggests that anthropogenic aerosols lowered the frequency of tropical storms in the last century.

2.2. Aerosol over the Mediterranean Sea

Over the Mediterranean Sea, aerosols have a unique atmospheric composition, with large spatial and temporal variability (Barnaba and Gobbi, 2004; Rodrıguez et al., 2002). Natural and anthropogenic aerosol is found in multiple layers and even in remote locations. Primary reasons are circulation patterns and geographic characteristics. An on-going international international research project in the Mediterranean area intends to improve the knowledge of the variability in the water cycle (Hydrological Cycle in Mediterranean Experiment (HyMeX), Drobinski et al., 2014). A related project has its focus on aerosol and chemistry (Chemistry-Aerosol Mediterranean Experiment (ChArMEx), Dulac et al., 2009).

The air flow over the Mediterranean is influenced by orography, causing blocking and channeling of air masses. The annual circulation pattern can be divided into a warm and a cold phase with little transition time (Maheras et al., 1999). The mean circulation in summer is assigned to the North Atlantic Ocean anticyclone (just barely extending over Europe) and the monsoon activity over the Indian Ocean. The mean circulation in winter is associated with intense cyclonic activity. In both periods the general wind direction within the lower troposphere is from north to south or south east with significant variations for different regions (Kallos et al., 2007).Differential heating of the European and North African land masses, with the sea in between, causes large spatial gradients and thermal circulations up to a few thousand kilometers (Millán et al., 2005).

In the warm season, the height of the mixing layer over Europe is approximately 1 to 2 km (Kallos et al., 2007) with temporal and regional variations. Over the Sahara Desert, the diurnal variation in the height of the mixing layer is larger than that over Europe, with heights reaching over 4 km in daytime (in the warm season). Over the Mediterranean Sea, there is almost no diurnal cycle of the 2 m temperature, and the annual variations in the height of mixing layer are quite small. The mixing layer height varies between 200 and 350 m. Islands and coasts may cause internal boundary layers and changes in the mixing depth. Convection and orographic flow can transport aerosols from the boundary layer to the free troposphere (Kallos et al., 2007).

Sources of anthropogenic aerosols are big cities, industrial activities, the energy sector, as well as the transport sector including land, sea, and air traffic. The stationary sources are mainly located on the European mainland with smaller contributions from North Africa and the sea (Lyamani et al., 2015).

A source of natural mineral dust particles above the Mediterranean is the Saharan Desert (Knippertz and Stuut, 2014). Mineral dust has a size range between 0.1 and 50 μm and an atmospheric turnover time of 3 to 7 days (Zender et al., 2004). This dust greatly contributes to the total particulate matter over the Mediterranean Sea (Rodríguez et al., 2001). Typical mass concentrations of mineral dust particles are 300 to 800 µg/m^3, with number concentrations of 1 to 100 particles/cm^3 between 1 and 4 km (Song and Carmichael, 1999).

The second source of natural aerosol (e.g., sea salt particles and dimethylsulfide (DMS)) is the Mediterranean Sea. The emission of sea salt particles mainly depends on the wind speed and two major processes: the bursting of air bubbles and the directly release from wave crests. Sea salt aerosols exist rather as solution droplets than as dry particles, although the droplets evaporate in the atmosphere. The typical diameters range between 0.05 and 100 µm, with an atmospheric turnover time of half a day (global average) (Textor et al., 2006). The typical mass concentrations of sea salt are 10 to 20 µg/m^3, with number concentrations of 100 to 300 particles/cm^3 below 1 km (Song and Carmichael, 1999).

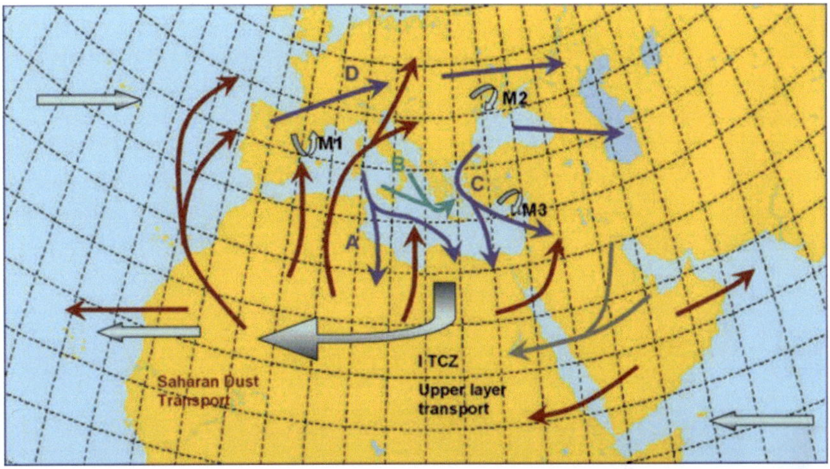

Figure 2.1.: Characteristic paths of pollutants and aerosols in the Euro-Mediterranean region. The blue and green arrows (A, B, C, and D) indicate the transport paths of anthropogenic pollutants in the lower troposphere. The grey arrows show transport paths of anthropogenic pollutants in the upper troposphere. The dark red arrows denote transport paths of mineral dust. M1, M2, and M3 indicate important thermal circulations. The paths undergo significant changes related to the synoptic-scale weather patterns in mid-latitudes. Taken from Kallos et al. (2007).

Figure 2.1 shows the characteristic transport paths of air masses and aerosols in the Mediterranean region (Kallos et al., 2007). The arrows display the most important pathways, which occur in different seasons, summarized over all seasons. Thus, the arrows do not represent averaged wind systems. The grey arrows display the anthropogenic pollutants in mostly upper-tropospheric layers whereas the blue and green arrows show the transport of the anthropogenic pollutants in the lower-tropospheric layers. The dark red arrows indicate the transport of mineral dust (also in the lower troposphere).

Kallos et al. (2007) summarized the characteristics from several studies over the last 20 years. Accordingly, air masses are transported during the whole year from southern continental Europe toward the Mediterranean Sea, Middle East and North Africa within 2 to 3 days (denoted by A, B, and C, respectively, in Figure 2.1). Thermal circulations at coasts (M1,

Kottmeier et al., 2000, M2, and M3), besides the long-range transport, are mainly responsible for the air quality. Over the Mediterranean Sea, transport of pollutants mainly occurs within the boundary layer, but islands can force polluted air masses to reach the free troposphere. In some regions (M1) vertical transport distributes pollutants over several layers. In coastal zones, the venting of industrial plumes toward the marine boundary layer can result in long travel distances of pollutants before they reach land again.

2.3. Medicane

Figure 2.2.: Satellite image of a Medicane on January 16, 1996 at 13 UTC at a wavelength of 10.8 µm. Taken from Fita et al. (2007).

Many cyclones impact the Mediterranean Sea (Pettersen, 1956). A few of these numerous cyclones are characteristic of tropical cyclones, i.e., a circular cloud pattern surrounding a cloudless eye (e.g., Billing et al., 1983; Ernst and Matson, 1983). These systems are called Medicanes, which is a neologism of Mediterranean hurricanes that describes the features and geographical connection. An example of a Medicane is displayed in Figure 2.2. The eye and the cyclonic cloud structure is clearly visible. Similar systems that exhibit tropical characteristics are polar lows. In the literature, polar lows are sometimes compared with Medicanes (Rasmussen and Zick, 1987).

Smaller vessels in the Mediterranean Sea are potentially endangered by Medicanes. Furthermore, coastal regions and islands in the Mediterranean may also suffer from Medicanes. When near the coast or making landfall, Medicanes can produce extensive damage. Medicanes often produce heavy precipitation, strong winds, high surfbreaks and flooding. To minimize damage, it is important to know where and when such systems occur. Forecasts should be very precise, particularly in terms of wind speed and precipitation.

Unstable conditions and deep convection are precursors for the development of Medicanes. The highest occurrence is in fall and winter, when the sea surface temperature is still warm compared with the overlaying air. Medicanes often develop under an upper-level temperature anomaly. In most cases, this anomaly is a cold cut-off low in the middle to upper troposphere (Homar et al., 2003). Recently, Chaboureau et al. (2012b) found jet crossing to be another trigger for Medicane development. Jet crossing is the passage of a low-level anomaly from one side of an upper-level jet to another. This can impact cyclogenesis because the upper-level forcing changes with the side of the jet. Fita et al. (2007) and Tous and Romero (2013) determined favorable conditions for the development of Medicanes; these conditions also hold for the development of hurricanes. The authors found that a high sea surface temperature leads to a thermodynamical imbalance between the air and the sea. This imbalance not only enforces but also accelerates cyclone formation. A high relative humidity occurs at the mid-levels, and high vorticity occurs at low levels. Moreover, large vertical wind shear between low and high levels hinders or prevents Medicane formation. The findings agree with the theoretical study by Emanuel (2005).

The evolution of a Medicane can be described in three phases (Luque et al., 2007). The first phase is called the pre-eye phase or initiation phase. A pre-existing vortex (formed, e.g., by the lee effect of mountains) and an upper-level anomaly are presuppositions for the evolution of a Medicane

(Chaboureau et al., 2012b). Because of the warm water and the cold air, there is a large air–sea thermodynamic disequilibrium, which is also a favorable condition for building polar lows (Emanuel, 2005). The vertical gradient in temperature (warm air in lower levels, cold air in upper levels) is a favorable condition for air rising from the lower levels and starting convection. Divergence in higher levels cause removing mass from lower levels, and pressure near the surface drops. The rising air cools while the relative humidity increases. Evaporation from the sea causes an increase in latent heat. These energy fluxes enhance convection. Heavy rainfall is produced due to the deep convection. Miglietta et al. (2013) described a high lightning frequency in this phase, which also occurs with hurricanes. The lightning strikes are located near the cyclone center.

The second phase is called the mature phase or stationary phase. In this phase, the convection organizes around a warm vortex. Because of the strong convection, the warm core is lifted to higher levels. This process weakens the vertical temperature gradient and the baroclinicity. When the upper cold core vanishes, the system has tropical characteristics. For the first time, an eye is visible and is surrounded by an axisymmetric cloud pattern (see Figure 2.2). While the system shows clear cyclonic rotation and strong winds, it travels rather slowly. Heavy precipitation can still occur, but with the increased wind speeds and deepening of the Medicane, the lightning frequency decreases and the strikes appear farther from the center compared with the pre-eye phase (Miglietta et al., 2013). The main active processes are the coupled interaction between deep convection and sea surface fluxes (Miglietta et al., 2011; Chaboureau et al., 2012b). Tous et al. (2013) found that the surface heat fluxes are the main factors controlling the intensity and track of Medicanes.

In the last phase, the so-called itinerant phase, the Medicane moves rapidly in a distinct direction. The wind speeds increase compared with the previous phases, but the rainfall decreases. When the Medicane makes landfall, it quickly dissipates because it lacks energy (the energy originates from the sea, Fita et al., 2007).

Cavicchia and von Storch (2012) provide an overview of the characteristics of Medicanes; a brief summary is provided here. A typical Medicane radius is between 70 and 200 km, and the distance traveled ranges from 700 to 3000 km over 12 h to 5 days. Wind speeds up to approximately 40 m/s are typical. The eye is visible for up to 3 days; in some cases, the eye develops twice, with an interruption as the Medicane crosses land areas, such as islands. The authors also found that the most favorable regions for the genesis of Medicanes are the Balearic Islands, the Ionian Sea and, lastly, the Aegean Sea (Figure 2.3). Autumn is the preferred season for Medicane occurrence, but some Medicanes have also been detected during winter and spring (Figure 2.4).

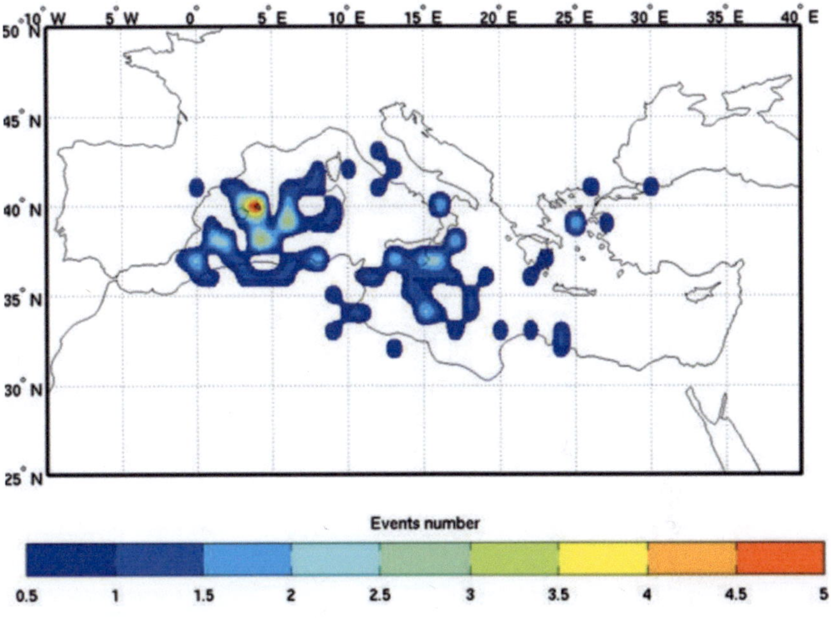

Figure 2.3.: Number of detected Medicanes events in a 1 ° × 1 ° grid spacing over 63 years. Taken from Cavicchia et al. (2014).

Medicanes evolve under environmental conditions that differ from those of hurricanes (Fita et al., 2007). The maritime tropical environment is characterized by more homogeneous atmospheric constitution. The air is warm and humid, with small seasonal variations. In contrast, the Mediterranean is characterized by a basin surrounded by complex orography. The air is influenced by air masses formed over neighboring continents of Europe and Africa and is often hot and dry. Thus, the potential for tropical cyclogenesis is generally small. However, Emanuel (2005) showed that the warm Mediterranean under a cut-off low at upper levels is a favorable condition for the generation of a Medicane. Ascending air cools and humidity increases. In combination with warm sea surface temperature, this area is an ideal incubator for a hurricane-like development. Sea surface temperatures of at least 26° C are necessary for the genesis of tropical hurricanes (Palmen, 1948), while Medicanes can form at much lower temperatures of around 15° C (Tous and Romero, 2013).

Deep convection and the strong influence of surface latent and sensible heat fluxes are characteristic of both types of systems (Emanuel, 2003).

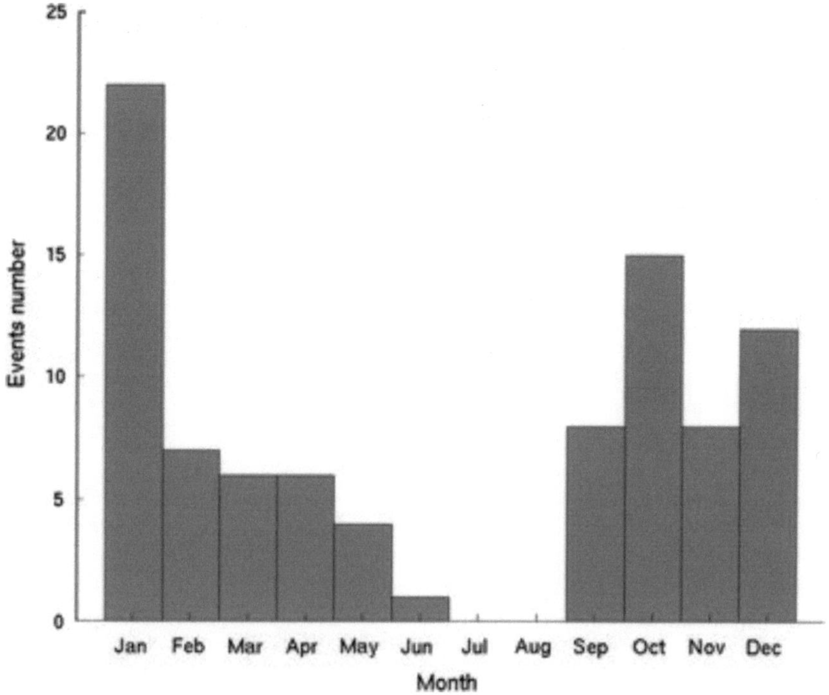

Figure 2.4.: Number of detected Medicane events per month, accumulated over 63 years. Taken from Cavicchia et al. (2014).

Also, the structures are similar, with an eye, eyewall and symmetrical cloud pattern.

Despite their similar structures, Medicanes are much smaller than tropical cyclones. In addition, the central pressure is higher and the horizontal winds are calmer in Medicanes, although the damage potential is still high. The lifetime of Medicanes is relatively short.

Because of their rarity, small size and short lifetime, Medicanes are not easy to detect or forecast. The lack of a dense and uniformly distributed measurement network over the sea makes detection even harder. When a Medicane does not make landfall, it may not be recognized at all. In recent years, a climatology of Medicanes and automatic detection algorithms have been developed.

Tous and Romero (2013) detected Medicanes in satellite data of Meteosat based on special criteria to enhance their predictive capability. The criteria for the detection is listed in Table 2.1. Conte et al. (2011) used numerical simulations and satellite images to study Medicanes by testing different instability parameters. Miglietta et al. (2013) included lightning data in their analysis to create an automatic detection algorithm. Fita et al. (2009) improved the simulation of two Medicanes (12 September 1996 and 18 October 2003) by assimilating precipitation from satellites and lightning data.

Cavicchia and von Storch (2012) and Cavicchia et al. (2014) used a dynamical downscaling approach with a regional climate model to create a climatology of Medicanes. To conduct downscaling, the authors spectrally nudge reanalysis data. Medicanes were detected based on the criteria listed in Table 2.1. Romero and Emanuel (2013) also used a climate model to create a climatology using a completely different technique. These authors used a statistical-deterministic approach to create thousands of synthetic storms. Walsh et al. (2014) used a regional climate model to estimate occurrence of Medicanes under climate change. The results indicate that the number of Medicanes will decrease because the environmental conditions favorable for genesis will be less frequent. Akhtar et al. (2014) coupled a regional climate model with an ocean model to successfully improve the simulation of historical Medicanes in terms of the agreement with observations.

Table 2.1.: Criteria to detect Medicanes in different studies

Tous and Romero (2013)	◦ "continuous cloud cover" ◦ symmetric cloud shape around eye ◦ diameter $< 300\,\mathrm{km}$ ◦ lifetime $\geq 6\,\mathrm{h}$ *timespan:* 1982 - 2003 *number of detected Medicanes:* 12
Picornell et al. (2014)	◦ $\Delta p \geq 3.2 \cdot 10^{-2}$ hPa/km in at least 6 of 8 principal directions ◦ thermal symmetry[1] with $B < 10$ m ◦ warm core structure[2] $-\mathbf{v_T}^L > 0\,\mathrm{m/s}$ and $-\mathbf{v_T}^U > 0\,\mathrm{m/s}$ *timespan:* tested on 4 Medicanes *number of detected Medicanes:* 3 out of 4
Cavicchia et al. (2014)	◦ thermal symmetry[3] with $B < 10$ m ◦ warm core structure[2] $-\mathbf{v_T}^L > 0\,\mathrm{m/s}$ and $-\mathbf{v_T}^U > 0\,\mathrm{m/s}$ ◦ $\overline{\mathbf{v_{r=50\,km}}} \leq 18$ m/s for at least 10 % or 6 h of track[4] ◦ $\overline{\mathbf{v_{r=50\,km}}}\vert_{850\,\mathrm{hPa}} \geq \overline{\mathbf{v_{r=50\,km}}}\vert_{300\,\mathrm{hPa}}$ ◦ $\max(\mathbf{v_{r=50\,km}}) \leq 29$ m/s for at least 4 h *timespan:* 1948 - 2011 *number of detected Medicanes:* 99

1 B is the storm-motion-relative lower troposphere thickness asymmetry, defined by $B = \overline{z_{700} - z_{925}}|_L - \overline{z_{700} - z_{925}}|_R$, with z_{700} means the height in 700 hPa and $|_L$ ($|_R$) indicates left (right) of storm motion; values are taken at a distance of 90 km left and right of storm center (adjusted from Hart (2003)).

2 $-\mathbf{v_T}$ being the thermal wind in the lower (L) and upper (U) troposphere (see also Hart (2003).)

3 basically the same as in [1] but with the original values of Hart (2003) for the thickness between 600 hPa and 900 hPa.

4 $\overline{\mathbf{v_{r=50\,km}}}$ is the average wind speed in around the pressure minimum with a radius of 50 km.

Numerical weather prediction models have also been applied for studying Medicanes. Homar et al. (2003) performed a case study of a Medicane that occurred on 12 September 1996 to evaluate the influence of various processes. A case study using a numerical weather prediction model of a Medicane that occurred on 26 September 2006 was performed by Miglietta et al. (2011). The authors found that a cold sea surface temperature delayed the development and reduced the intensity of the Medicane. Miglietta et al. (2014) investigated the influence of various cloud micro-physics schemes on the simulation of the same Medicane. The authors found that simple schemes (without taking graupel into account) reproduced this particular Medicane better than more complex (two-moment) micro-physics schemes. Picornell et al. (2014) developed criteria to identify Medicanes in numerical weather prediction output; these criteria are listed in Table 2.1. The studies mentioned above by Conte et al. (2011) and Miglietta et al. (2013) were also performed with numerical weather prediction models.

Tous and Romero (2013) used very strict criteria to detect Medicanes, which statistically occurred every two years between 1982 and 2003. The work of Cavicchia and von Storch (2012) revealed a frequency of approximately 1.6 Medicanes per year between 1948 and 2011. An informal listing found on the website of Grup de Meteorologia of the Universitat de les Iles Balears (Grup de Meteorologia UIB, 2007) shows approximately 1.5 Medicanes per year since 1982 (Romero and Emanuel, 2013). Assuming that 2 Medicanes occur per year, Romero and Emanuel (2013) found that in the future, the number of Medicanes may decrease, but the intensity and damage potential may increase. Cavicchia et al. (2014) did not find a significant trend in the occurrence of Medicanes, although the year-to-year variability was high. The authors did not expect a more frequent occurrence of Medicanes in the future. Walsh et al. (2014) agreed with these two studies as they found that the number of Medicanes may decrease.

3. Model System COSMO-ART

This chapter presents the model system used in this research. Information that has been described in previous studies (with remarks linking to those descriptions) is only briefly discussed, while those components that have not been previously discussed and that are important for this work, with the particular goal of short- to medium-range weather prediction, are described in great detail.

3.1. COSMO and Specific Features

The **CO**nsortium for **S**mall-scale **MO**deling (COSMO) is an association of national meteorological services and military and regional services. COSMO developed a limited area model to predict atmospheric states (Baldauf et al., 2011). The COSMO-Model is used for numerical weather prediction and for atmospheric and climate research (COSMO-CLM).

Germany's national meteorological service (**D**eutscher **W**etter**D**ienst, DWD) uses the COSMO-Model in two configurations for numerical weather prediction. COSMO-EU covers the entirety of Europe with a horizontal grid spacing of 7 km and 40 non-equidistant vertical layers. The predictions are used, among other things, as boundary conditions for the second configuration, COSMO-DE, which is the core of weather prediction by DWD in Germany. COSMO-DE covers Germany and neighboring regions with a horizontal grid spacing of 2.8 km and 50 vertical layers. To produce a reliable forecast, both models assimilate data from observations.

Within the model, the primitive hydro-thermodynamic equations that describe fully compressible, non-hydrostatic flow in a moist atmosphere are solved (Schättler et al., 2011). Horizontal geographical coordinates (latitude and longitude) and a generalized terrain-following vertical coordinate compose the coordinate system. The dedicated grid structure is a horizontal Arakawa C-grid and a Lorenz staggered vertical grid.

A grid structure based on geographical coordinates has two major disadvantages. First, meridians converge at the poles, leading to numerical singularities. Second, the distance between two grid points depends on the latitude; thus, grid cells have different sized base areas. To overcome these disadvantages, the North Pole is rotated such that the simulation area covers the equator.

Spatial discretization is achieved with finite differences of an arbitrary order (second order is default). For the temporal integration, a default Runge-Kutta scheme is used, but other schemes are also available. For numerical efficiency, the parameterizations that describe sub-grid scale processes are split into slow and fast components and are solved at individual time steps.

3.1.1. Application of Spectral Nudging

To run a regional model, initial and boundary conditions are needed to force the model. Usually, global models provide these data. Nested runs use data from a coarse horizontal model grid with a large horizontal extension. Ideally, one can see the same large-scale features in all runs, but in truth, these features are sometimes displaced. Spectral nudging is one technique in atmospheric modeling to keep the solution of a model run close to the driving model. Spectral nudging is often applied to tropical cyclones (Feser and Barcikowska, 2012), particularly to represent the track of a cyclone properly.

The concept of spectral nudging is to pull the interior of a model domain closer to the driving model, similar to the treatment of the boundary fields. The interior of a model domain includes everything except the zone in which the driving model contributes to the boundary conditions. The solution of the model interior for a meteorological variable Ψ is adjusted by the coarse model state Ψ^c

$$\Psi^n = \eta \cdot \left(\Psi^c - \Psi\right). \tag{3.1}$$

This equation depends on the nudging coefficient η. The nudged state of the model Ψ^n replaces the original solution Ψ. In COSMO, spectral nudging is implemented as described by von Storch et al. (2000).

Because the large-scale phenomena should only be accounted for by nudging, the variable that is nudged is Fourier transformed in spectral space. Nudging can then be applied to spatial scales larger than a chosen value, which is determined by the wave number. The expansion of a two-dimensional variable in time is

$$\Psi(\lambda, \phi, t) = \sum_{j=-N_j}^{N_j} \sum_{k=-N_k}^{N_k} \psi_{j,k}(t) \exp\left(\frac{\imath j \lambda}{L_\lambda}\right) \exp\left(\frac{\imath k \phi}{L_\phi}\right). \tag{3.2}$$

j and k are the wave numbers in the zonal and meridional directions. λ and ϕ are the zonal and meridional coordinates. L_λ is the extension in the zonal direction and, analogously, L_ϕ is the extension in the meridional direction. N_j is the number of zonal wave numbers, and N_k is the number of meridional wave numbers. $\psi_{j,k}$ is the coefficient of the Fourier transformation and is dependent on time t.

Analogically, the expansion of a chosen two-dimensional variable of the coarser model is described by

$$\Psi^c(\lambda, \phi, t) = \sum_{j=-N_j^c}^{N_j^c} \sum_{k=-N_k^c}^{N_k^c} \psi_{j,k}^c(t) \exp\left(\frac{\imath j \lambda}{L_\lambda}\right) \exp\left(\frac{\imath k \phi}{L_\phi}\right). \tag{3.3}$$

For the number of wave numbers, the conditions $N_j^c < N_j$ and $N_k^c < N_k$ are fulfilled.

The nudging coefficient η in Equation 3.1 provides the strength of the nudging. Nudging shall only affect large-scale conditions, which are driven through processes at high altitudes, i.e., at higher levels of the model. Therefore, nudging depends on the height and respective pressure, and η decreases with increasing pressure (decreasing height). Above a specific pressure value p_n (below a specific height), which can be chosen by the modeler, nudging ceases. Thus, the model can still respond to small-scale effects. The pressure-dependent nudging coefficient η^0 is given by

$$\eta^0(p) = \begin{cases} \alpha\left(1 - \frac{p}{p_n}\right)^2 & \text{for } p < p_n, \\ 0 & \text{for } p > p_n. \end{cases} \tag{3.4}$$

The maximum wave numbers j_n and k_n, to which nudging is applied, must be chosen by the modeler. All features within the lower wave numbers are nudged, whereas features within the higher wave numbers only respond to the model. The nudging coefficient in a given pressure level is then

$$\eta_{j,k} = \begin{cases} \eta^0(p) & \text{for } j \leq j_n \text{ and } k \leq k_n, \\ 0 & \text{for } j > j_n \text{ and } k > k_n. \end{cases} \tag{3.5}$$

With Equation 3.2 and Equation 3.3 and the nudging coefficient of Equation 3.5 the nudging equation (Equation 3.1) can be written as

$$\Psi^n(\lambda, \phi, t) = \sum_{j=-N_j^c}^{N_j^c} \sum_{k=-N_k^c}^{N_k^c} \eta_{j,k} \left[\psi_{j,k}^c(t) - \psi_{j,k}(t)\right] \exp\left(\frac{\imath j \lambda}{L_\lambda}\right) \exp\left(\frac{\imath k \phi}{L_\phi}\right). \tag{3.6}$$

In the coarse run (see later in section 4.2), spectral nudging is performed to drive the GME model, with $\alpha = 0.05$ and a limiting pressure level of $p_n = 850$ hPa (Equation 3.4). In this scenario, only the horizontal wind components u and v are nudged. The maximum wave numbers for nudging are j_n and $k_n = 4$, which corresponds to a spatial scale of approximately 112 km (Equation 3.5). Thus, all features larger than 112 km are nudged. Nudging is applied every three hours, which is the same time step used for updating variables at the boundary by the driving model.

3.1.2. Generation of Synthetic Satellite Images

Comparing satellite data with model output is challenging because satellites measure the brightness temperature, which is not directly simulated by COSMO. One possibility for such a comparison is to calculate comparable quantities of COSMO variables (the "observation-to-model approach"). To retrieve cloud information, assumptions must be made, and complex algorithms are applied. The result is many uncertainties and errors. Another possibility is a so-called "model-to-observation approach", where the measurements of satellites are simulated using model variables (e.g., Morcrette, 1991; Chaboureau et al., 2012a). These data can be directly compared with measurements by the satellite.

COSMO can follow the model-to-observation approach by producing synthetic satellite images. This process is realized by the coupling of the fast radiative transfer model for the Television Infrared Observation Satellite (TIROS) Operational Vertical Sounder (RTTOV) (Keil et al., 2006). RTTOV was developed by ECMWF (Eyre, 1991) and improved by EUMETSAT. RTTOV is a numerical model that simulates the brightness temperature for satellite infrared and microwave radiometers. For input data, the model requires profiles of temperature, specific humidity, cloud liquid and ice water, water content of snow, cloud cover and pressure. Additionally, two-dimensional fields are needed: surface pressure, skin temperature, 2-meter temperature, 2-meter specific humidity, and the land-sea mask. The output variables are brightness temperature and clear- and cloudy-sky radiances that correspond to the measured variables by the chosen satellite channels (infrared and water vapor).

3.2. Two-Moment Cloud Micro-Physics Scheme

COSMO uses a bulk water-continuity scheme to represent clouds (Doms et al., 2011); this scheme is a compromise between computational cost and the accurate description of physical processes. Bulk models cluster hydrometeors into several groups and directly predict the mass fraction to limit the number of equations needed. An assumption of the size distribution for the hydrometeors must be made; thus, the shape of the distribution depends on the predicted moments in general. Compared with this scheme, spectral models are more accurate. These models subdivide the hydrometeor sizes into many categories and allow the direct application of multiple micro-physical principles. Spectral models are very complex, and current computer capacities prevent their use in numerical weather prediction.

In the default micro-physics scheme of COSMO, the size distribution of the hydrometeors only depends on the mass fraction q^x, where x represents each of the hydrometeors: cloud droplets c, rain drops r, cloud ice i, snow s, and graupel g. The mass fraction is proportional to the third moment. There is also the possibility of using a two-moment micro-physics scheme (Seifert and Beheng, 2006), in which the size distribution and physical processes depend on the mass fraction q^x and on the number density N^x of the particles. Because the two-moment micro-physics schemes outperform single-moment schemes in general, the use of two-moment schemes is highly recommended (Igel et al., 2015). Of course, the computational cost increases, but two-moment schemes are not as expensive as a spectral model. The two-moment scheme includes an additional hydrometeor type: hail h.

The budget equations for the number and mass concentration of hydrometeor class x in flux form is

$$\partial_t N^x = -\nabla \cdot (\mathbf{v} N^x) + \nabla \cdot \mathbf{F}_{N^x} + \partial_z \left(\overline{w_x^{sed} N^x} \right) + S_{N^x}^{\mathrm{act}} + S_{N^x}, \quad (3.7)$$

$$\partial_t m^x = -\nabla \cdot (\mathbf{v} m^x) + \nabla \cdot \mathbf{F}_{m^x} + \partial_z \left(\overline{w_x^{sed} m^x} \right) + S_{m^x}^{\mathrm{act}} + S_{m^x}, \quad (3.8)$$

$$x = c, r, i, s, g, h.$$

On the right hand side of Equation 3.7 (Equation 3.8), the first term denotes changes in the number (mass) concentration due to advection, the second term denotes changes due to turbulent diffusion, the third term denotes changes due to gravitational sedimentation (for all hydrometeors but cloud droplets), the fourth term denotes changes due to activation of aerosol particles (for cloud droplets and ice; otherwise, the term is zero), and the last term denotes changes due to micro-physical processes. These micro-physical processes differ by hydrometeor type. For example, for cloud droplets, these processes include the self-collection of cloud droplets (the number changes while the mass concentration remains the same), the formation of raindrops by collision and coalescence (also called auto-conversion), the accretion by raindrops, the collection by ice, snow, graupel, and hail (also called riming), the melting of ice crystals, the freezing of cloud droplets, and the conversion to graupel or hail. Seifert and Beheng (2006) list processes for the other hydrometeor types.

3.3. COSMO-ART

ART ('**A**erosols and **R**eactive **T**race gases") (Vogel et al., 2009) is an extension of COSMO that describes relevant regional-scale processes for gases and liquid and solid particles in the atmosphere. ART is coupled to COSMO online; thus, meteorological variables, e.g., temperature, that are delivered by COSMO are not constant in ART but are alterable by feedback processes. COSMO uses the updated values for further calculations. Furthermore, online coupling allows for a consistent description of comparable processes (i.e., advection) because the same numerical methods and grid representations are used. COSMO-ART has been used for several case studies and validations (i.e., Knote et al., 2011; Bangert et al., 2011; Athanasopoulou et al., 2013; Rieger et al., 2014; Baklanov et al., 2014). Figure 3.1 gives an overview of the feedback process considered by COSMO-ART.

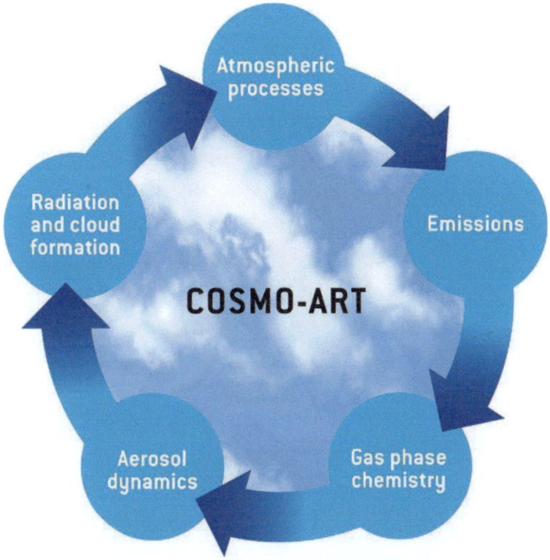

Figure 3.1.: Chain of feedback processes in COSMO-ART.

The Regional Acid Deposition Model Version Karlsruhe (RADMKA, Vogel et al., 2009) mechanism is included in COSMO-ART to represent tropospheric gas-phase chemistry. RADMKA is based on the Second Generation Regional Acid Deposition Model (RADM2, Stockwell et al., 1990) and is extended by accounting for reactions of isoprene (Geiger et al., 2003) and monoterpens (Stockwell et al., 1997). The photolysis rates are fully coupled to the radiation scheme used in COSMO (GRAALS, or General

Radiative Algorithm Adapted to Linear-Type Solutions, Ritter and Geleyn, 1992). The corresponding differential equations can be solved by the Kinetic PreProcessor (KPP, Damian et al., 2002) or by an explicit solver. There is an option to consider gas-phase chemistry of halogen compounds (Kraut, 2011).

The rates between gaseous ammonia and ammonium as aerosols, sulfuric acid and sulfate, nitric acid and nitrate, and water are calculated by version 2 of ISORROPIA (Fountoukis and Nenes, 2007). The rates of the volatile organic compounds of gas and aerosols are calculated by a volatility basis set (VBS, Athanasopoulou et al., 2013).

3.3.1. Representation of Aerosols

Within COSMO-ART, the size distributions of aerosols are represented by modes. Therefore, the number and mass of each mode are described by log-normal distributions based on the corresponding moments. Comparable to the representation of hydrometeors in section 3.2, finely tuned size bins precisely cover a measured distribution, but the computational cost is too high. A modal approach is a good balance between the computational burden and process details (Tao et al., 2012).

There are twelve overlapping modes in COSMO-ART, which are described by the extended module called Modal Aerosol Dynamics Model for Europe extended by Soot (MADEsoot, Riemer et al., 2003; Vogel et al., 2009). The modes are further specified in Table 3.1. Each of the three modes specify mineral dust ($dustA$, $dustB$, $dustC$) and sea salt aerosols ($seasA$, $seasB$, $seasC$). The coarse mode (c) consists only of anthropogenic aerosols with a maximum diameter of 10 µm (also called PM10). The dust and coarse modes do not interact with the gas phase or the other modes, whereas the sea salt mode interacts with the gas phase but not directly with the other modes. There is one pure soot mode (s), two mixed modes without soot at different size ranges, nucleation (if) and accumulation mode (jf), and two mixed modes that include soot in the same size ranges (ic and jc). The composition of these five modes changes over time by nucleation from the gas phase and coagulation. Two other aerosol modes are available but are treated in a different way. These modes represent pollen and volcanic ash. In this study, pollen and volcanic ash are not accounted for.

For mode i, the corresponding size distribution of the number density n_i as a function of the particle diameter d_p is given by

$$n_i(d_p) = \frac{N_i}{\sqrt{2\pi} \cdot \ln \sigma_i \cdot d_p} \exp\left(- \frac{\left(\ln d_p - \ln \overline{d_{p,N,i}}\right)^2}{2 \cdot \ln^2 \sigma_i} \right). \tag{3.9}$$

N_i is the total number concentration of aerosol particles, σ_i is the geometrical standard deviation, and $\overline{d_{p,N,i}}$ the median diameter of the particles with respect to the number distribution, each referring to mode i. Figure 3.2 shows the size distribution for the number concentrations of all modes included in COSMO-ART. The calculation of the curves in Figure 3.2 uses the initial values of σ_i and $\overline{d_{p,N,i}}$ within COSMO-ART. These values and the values for N_i, which are chosen to build a hypothetically environment over Mediterranean coastal regions, are listed in Appendix A. The standard deviation for the number and mass distribution is constant.

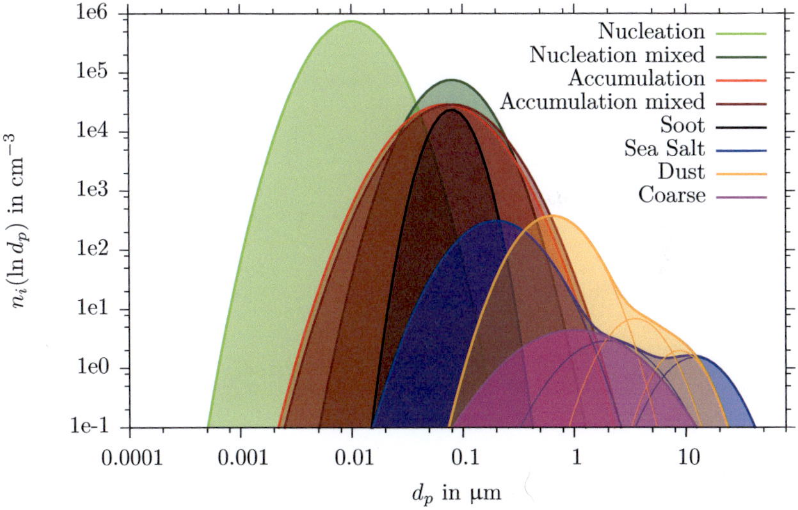

Figure 3.2.: Size distribution for number concentrations of all modes in COSMO-ART.

Table 3.1 provides an overview of the considered modes in COSMO-ART, including their abbreviated names, chemical compositions (sulfate, nitrate, ammonium, organics, or soot) and size affiliation. Sea salt is assumed to comprise sodium chloride, sodium sulfate and liquid water (Lundgren et al., 2013). Because mineral dust does not interact with the gas phase nor with the other modes, its chemical composition is of no interest here (Stanelle et al., 2010).

Table 3.1.: Description of the individual modes of the aerosol particles

Name	Chemical Composition
if	secondary particles (SO42-, NO3-, NH4+, organics, H2O)
jf	secondary particles (SO42-, NO3-, NH4+, organics, H2O)
s	pure soot
ic	aged soot (SO42-, NO3-, NH4+, organics, H2O, soot)
jc	aged soot (SO42-, NO3-, NH4+, organics, H2O, soot)
c	anthropogenic directly emitted particles (PM10)
$seasA$	Na+, Cl-, SO42-, HSO4-, H2+
$seasB$	Na+, Cl-, SO42-, HSO4-, H2+
$seasC$	Na+, Cl-, SO42-, HSO4-, H2+
$dustA$	mineral dust
$dustB$	mineral dust
$dustC$	mineral dust

The temporal change in the total number density of mode i is described by

$$\partial_t N_i = -\nabla \cdot (\mathbf{v} N_i) + \nabla \cdot \mathbf{F}_{N_i} + \partial_z \left(\overline{w_i^{sed} N_i} \right) + S_{N_i} + Q_{N_i}. \qquad (3.10)$$

On the right side of Equation 3.10, the first term denotes changes in the number concentration due to advection, the second term denotes changes due to turbulent diffusion, the third term denotes changes due to gravitational sedimentation, the fourth term denotes changes due to micro-physical processes and the last term represents source terms due to emissions (see subsection 3.3.2).

These micro-physical processes are nucleation from the gas phase, washout (scavenging of particles by precipitation), coagulation within each mode (this changes the total number of particles but not the total mass within each mode), coagulation of different modes, condensation (only changes the total mass), and transport due to sub-grid-scale convection. For mineral dust and the coarse mode (see above), only washout and convective transport remains. A detailed overview of these processes is given in Vogel et al. (2009).

3.3.2. Treatment of Gaseous and Aerosol Emissions

Before aerosol and trace gases are able to influence the environment, they must reach the atmosphere. This process is called emission and is denoted as Q_{N_i} in Equation 3.10. The emission of natural aerosols, such as mineral dust, sea salt particles, and biogenic emissions (isoprene and monoterpenes), is parameterized and calculated based on the atmospheric state. Other emissions, such as anthropogenic emissions, emissions due to biomass burning, and emissions of dimethyl sulfide (DMS), are expressed using external datasets. The following paragraphs present a short overview of the description of the most important emissions for this research.

Anthropogenic Emissions

Emissions of anthropogenic aerosols and gases are determined by an external data set. Such a dataset contains the annual emissions by pollutant, country, and source category (e.g., combustion, road transport, waste treatment, or agriculture). Time functions are applied to calculate emissions at a particular date and time. Emissions are further described by the spatial distribution of their source, which can be a point source (like a power station), a line source (roads and trails) or an area source (fields). This information is transferred into the grid structure of COSMO-ART.

Mineral Dust

Within COSMO-ART, there are several parameterizations for describing the emission of mineral dust. However, the basic concept of the process is the same in all parameterizations, as follows. A more detailed description of the parameterization can be found in Vogel et al. (2006).

The emission of mineral dust particles depends on the friction velocity. Large dust particles are more easily lifted than small particles because they have lower binding energies (cohesion forces). Because of gravitational forces, large particles do not remain in the air for very long; instead, these particles fall back down and release small particles during the ballistic flight. This process is called saltation bombardment. The small particles have a small settling velocity such that they can remain in the atmosphere for a long time and can travel great distances.

Two particles fluxes are needed to describe the process of saltation: the so-called horizontal saltation flux, which describes the flux of large particles near the surface, and the vertical emission flux of the small particles. In COSMO-ART, the horizontal flux is parameterized by an empirical function (White, 1979) that depends on the soil type, the number and size of particles on soil, and the critical friction velocity, which is the minimum value of the friction velocity required to move particles. This critical friction velocity also depends on the particle diameters, the humidity of the soils and the soil roughness length (Shao and Lu, 2000). Using the horizontal flux, the vertical dust flux can be determined (Alfaro and Gomes, 2001).

Sea Salt

The emission of sea salt particles mainly depends on the horizontal wind speed near the surface; it can be described by two mechanisms. Small drops (film and jet drops) are emitted by the bursting of air bubbles when air is mixed into water by breaking waves. For this purpose, wind speeds of at least 3 m/s are required. Large drops (spume drops) are emitted by mechanical tearing directly off of the wave crests by wind. Wind speeds larger than 9 m/s are required for this type of emission.

Within COSMO-ART, dry sea salt particles are assumed to be emitted, depending on the horizontal wind speed at a 10 m height. The impact of salinity on emissions is neglected. The impact of sea surface temperature on emissions is only taken into account for the smallest sea salt mode. Freshly emitted particles are assumed to consist of sodium chloride.

Empirical parameterizations are used to describe the emission of sea salt particles within COSMO-ART. Each parameterization results in a different size distribution. Sea salt mode A is represented by the parameterization of Mårtensson et al. (2003), mode B by Monahan et al. (1986), and mode C by Smith et al. (1993). A more detailed description of the processes and parameterizations can be found in Lundgren et al. (2013).

3.3.3. Representation of Interactions between Aerosols, Radiation, and Clouds

Aerosols alter the energy budget and influence cloud formation and hence impact precipitation and the water cycle. Aerosols directly scatter and absorb radiation and indirectly modify cloud physical properties as cloud condensation or ice nuclei. As a consequence, the hydrometeor size distributions may change.

Micro-physical processes of clouds and aerosols are coupled by the processes of activation and ice nucleation. In numerical weather prediction, a constant value of cloud condensation nuclei and ice nuclei is usually prescribed. Within COSMO-ART, an explicit description of aerosols provides an opportunity to describe the cloud-process modification by aerosols realistically (Bangert et al., 2012).

Activation

An aerosol particle grows by condensation of water vapor and forms a cloud droplet if the supersaturation is above a critical threshold. Supersaturation is the share of the water vapor mixing ratio that exceeds saturation. Köhler theory describes this process based on thermodynamic equilibrium by combining the Kelvin effect, which describes changes in water vapor pressure due to a curved surface, and Raoult's law, which describes changes in water vapor pressure due to a chemical solution.

According to Köhler theory, the growth of aerosol particles depends on the chemical composition of the solute and the atmospheric state. If the supersaturation is larger than zero and the droplet exceeds a critical diameter, then the droplet can continue to grow to raindrop size. If the droplet's diameter is below the critical value or if the supersaturation is below zero, then the droplet shrinks.

When the critical diameter is exceeded for a supersaturation larger than zero, the aerosol droplet is activated because it quickly grows to the size of a cloud droplet. Within COSMO-ART, an aerosol particle becomes activated when the maximum supersaturation in a cloud exceeds the critical supersaturation. Because there is no analytical method to solve this problem, parameterizations are used (Fountoukis and Nenes, 2005; Barahona et al., 2010b).

Mineral dust, which is assumed to consist of insoluble particles, is treated differently because physical adsorption affects the activation process. In this case, Frenkel, Halsey, and Hill (FHH) activation theory (Kumar et al., 2009) is used to describe the activation of aerosol particles to grow to cloud droplets. This theory results in a numerical solution for the critical supersaturation, which is achieved by the help of empirical fitting constants.

Ice Nucleation

The formation of ice crystals is based on two competing processes: heterogeneous and homogeneous freezing. If only water initiates the freezing process, then homogeneous freezing occurs. Homogeneous freezing occurs at temperatures below 235 K; in COSMO-ART, this process is parameterized by Barahona et al. (2010a).

Heterogeneous freezing occurs if aerosol particles are included in the freezing process. This process starts at higher temperatures than homogeneous freezing (approximately 263 K). In COSMO-ART, heterogeneous freezing is parameterized by Phillips et al. (2008). Here, the efficiency of freezing is described by aerosol composition, surface areas, and diameters. If only heterogeneous freezing occurs, then a simplified maximum supersaturation over ice is used, as parameterized by Barahona and Nenes (2009).

There is a specific number concentration of nucleated ice crystals that inhibits homogeneous nucleation (N_{lim}) because the maximum supersaturation over ice is lower than the supersaturation needed for homogeneous freezing. If the number concentration of nucleated ice crystals by heterogeneous freezing (N_{het}) exceeds N_{lim} or if the temperature is above 235 K, then only heterogeneous freezing occurs. In this case, the total number of nucleated ice crystals is equal to N_{het}. Otherwise, the number of nucleated ice crystals by homogeneous freezing N_{hom} also contributes to the total number of nucleated ice crystals.

Cloud Optical Properties

Aerosols directly scatter and absorb radiation, but they also indirectly affect radiation by changing cloud optical properties. To study this impact, the various droplet and ice crystal sizes must be taken into account in the radiation scheme of COSMO (GRAALS, Ritter and Geleyn, 1992). Thus, the extinction coefficient, single-scattering albedo and asymmetry factor of clouds are needed.

Because the explicit solution of the Maxwell equations is costly in terms of computations, parameterizations are used. Optical properties are described by Hu and Stamnes (1993) depending on the effective radii, mass concentration of cloud water and the wavelength. Because of the complex shape of ice crystals, the calculation of optical properties of ice clouds is complicated. In the radiation calculation, precipitating hydrometeors are not considered.

4. General Description: Case Study Medicane "Rolf"

Following the general overview of the aerosol-cloud-precipitation-interactions, the remaining part of this thesis focuses on these interactions in case of a particular Medicane. Because a Medicane leads to high emission of sea salt particles, these interactions are expected to be strong. In this chapter, the particular Medicane and the simulation of this Medicane is presented.

4.1. Synoptic Weather Situation

On November 3, 2011, an upper-level trough was situated over the eastern Atlantic Ocean. The trough shifted eastward and reached Europe the next day. As it shifted, the trough intensified and extended southward, covering the Mediterranean Sea.

In front of the trough a surface low developed during November 5, 2011, in the lee of the Pyrenees. This surface low was called "Rolf" by FU Berlin. The surface pressure was about 1000 hPa on this day. Rolf moved south-eastward.

Upon reaching the coast near Barcelona, Spain, on November 6, 2011, the upper-level trough had nearly disappeared. A cold upper cut-off low remained. Rolf didn't move to the east. Analysis of the surface pressure by DWD (Behrendt and Mahlke, 2011) shows the location of Rolf at 00 UTC of November 6, 2011, at the French coast (Figure 4.1(a)).

Rolf moved further south and reached the Balearic Islands on November 7, 2011. Rolf was then redirected toward Sardinia as it intensified via the

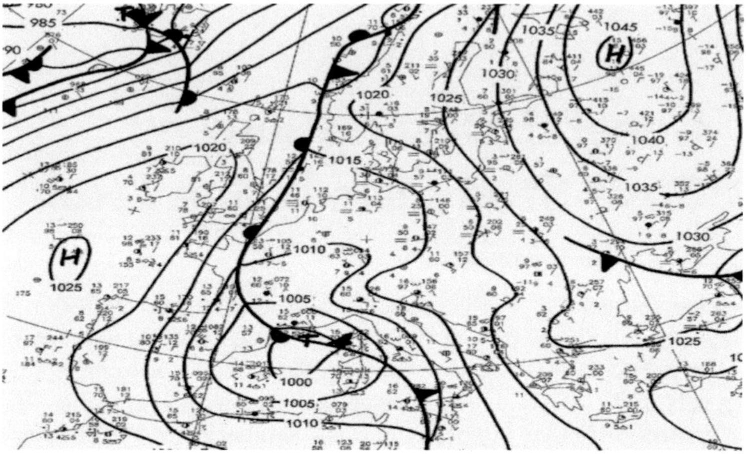

(a)

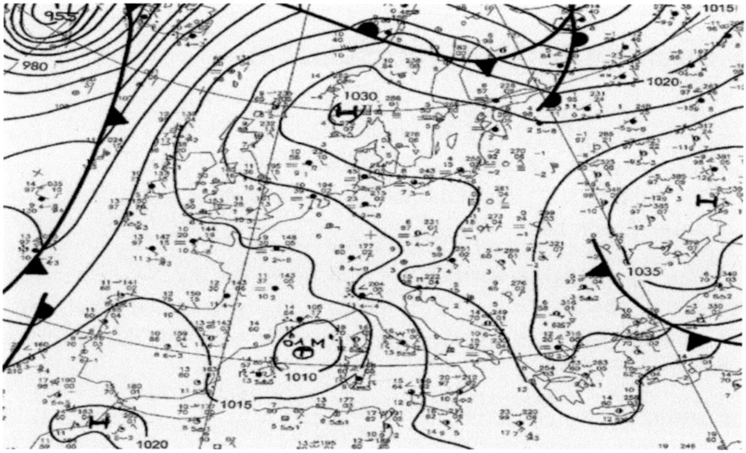

(b)

Figure 4.1.: Mean sea level pressure at 00 UTC by DWD for (a) November 6, 2011, and (b) November 8, 2011, both taken from Behrendt and Mahlke (2011).

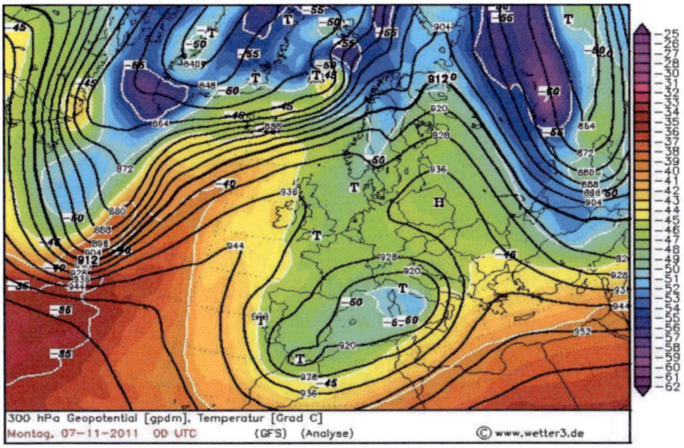

(a)

(b)

Figure 4.2.: (a) Temperature at 300 hPa by GFS at 00 UTC November 7, 2011. Taken from Behrendt and Mahlke (2011). (b) Visible satellite image at 10 UTC for November 8, 2011, taken from The Meteo Company B.V. (2011).

latent heat fluxes from the sea surface, which was approximately 22 ° C. The upper cold cut-off low was quasi-stationary and thus stayed above the surface low. An analysis of the Global Forecast System (GFS) by the National Centers for Environmental Prediction (NCEP) shows temperatures of approximately -50 ° C at 300 hPa (Figure 4.2(a)). Thus, a high vertical temperature gradient and strong convection with thunderstorms developed. At 18 UTC, the DWD classified the system as a "Medicane". Additionally, the Satellite Service Division of the US National Oceanic and Atmospheric Administration (NOAA) identified this system as a "tropical" cyclone with the identifier "01M" (see Table 4.1). 01 represents the number of registered tropical cyclones in the specific region ("M" for the Mediterranean Sea). Therefore, this system was the first cyclone with tropical characteristics that developed in this area in 2011.

Table 4.1.: Shortened Tropical Cyclone Information Bulletin on cyclone 01M by NOAA. Pressure is given in hPa and wind speed in knots.

Date	Time	Basin	Lat	Lon	Pressure	Wind speed	ID
20111107	1200	MED	40.6	-5.9	0	27	IN1
20111107	1800	MED	41.1	-5.3	997	35	01M
20111108	0000	MED	41.4	-5.4	991	45	01M
20111108	0600	MED	41.5	-5.8	991	45	01M
20111108	1200	MED	42.1	-6.2	991	45	01M
20111108	1800	MED	42.8	-6.5	991	45	01M
20111109	0000	MED	42.6	-6.4	991	45	01M
20111109	0600	MED	42.7	-6.1	1000	30	01M
20111109	1200	MED	43.2	-4.7	1006	25	01M

On November 8, 2011, Rolf again changed direction and headed toward the French coast. A minimum pressure of 991 hPa was determined by NOAA. The diameter of the Medicane was approximately 300 to 400 km. The sea surface had cooled to temperatures of approximately 17 ° C. In Figure 4.1(b), the location of the Medicane between the Balearic Islands and Corsica can be seen on the analysis of the surface pressure by the DWD at 00 UTC on November 8, 2011. The satellite image of the visible channel, Figure 4.2(b), shows the eye and the roughly axis-symmetric cloud pattern. These features are typical of a tropical cyclone.

On November 9, 2011, the Medicane made landfall in France. Because the source of energy (the latent heat fluxes of the Mediterranean Sea) was lost, the Medicane quickly weakened over land. But there were still disastrous consequences. High surf breaks occurred along the coasts of France and Italy. Heavy precipitation affected southern France and Switzerland and northwestern Italy. Locally, the precipitation totals over 24 hours were very high (465 mm in Vicomorasso and 350 mm in Crocetta di Orero, both in Genoa, Italy). Flooding and rivers overtopping their banks caused major damage by sweeping away trees and cars. A wind speed of approximately 150 km/h was measured at several locations. Overall, seven people were killed (all numbers are taken from Mühr, 2011).

4.2. Simulation Setup

The situation of Medicane Rolf (section 4.1) is simulated using the COSMO-ART model system (see description in chapter 3). The simulation spans November 2 to November 11, 2011. Triple nesting is used, in which the simulation area and the grid size are downscaled for each domain-specific configuration.

The simulation areas are plotted in Figure 4.3. The coarse domain covers the entire area plotted. The red quadrangles represent the borders of each nested domain. Table 4.2 provides an overview of the domain specific settings for the four areas (the number of horizontal grid cells are given in longitudinal grid cells $\times$ latitudinal grid cells). These configurations are the basis for several scenarios. Each of the configurations for the coarse, nest 1, and nest 2 domains is used to build one scenario; the main differences are the coverage and grid mesh size. The configuration of nest 3 is the basis for several scenarios and ensembles, which are described in greater detail later in this chapter. An overview of all scenarios can be found in Table 4.3.

A 3rd order Runge-Kutta scheme based on two time levels accomplishes the temporal integration. Horizontal advection for scalars (i.e., water constituents, aerosol and gases) is accomplished by a direction-split finite volume scheme (Bott, 1989). The configurations remain as close as possible to the default configurations of the operational forecasts of COSMO-

EU and COSMO-DE by the DWD, particularly for the domain-specific configurations of nests 2 and 3 because the grid mesh sizes are the same as those for COSMO-EU and COSMO-DE, respectively. Important differences and features are highlighted in the paragraphs below.

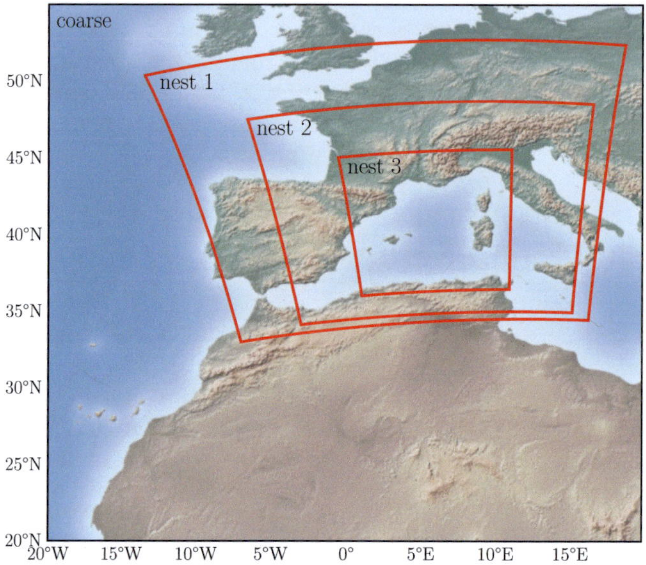

Figure 4.3.: Domains of the different configurations. The largest one is the coarse domain, the red quadrangles are the boarders to nest one to three (smallest).

The coarse-domain scenario using an unrotated grid is performed continuously from November 2 to November 11, 2011. The area is explicitly chosen to cover the Saharan Desert, which is the major source of dust particles over the Mediterranean Sea. Extending the domain southward allows dust particles to enter the area of interest. For the initial and boundary conditions, data from the global model of DWD, GME, is used. Additionally, spectral nudging is performed as described in subsection 3.1.1 for the horizontal wind components u and v. The boundary data and spectral nudging are updated every three hours.

Table 4.2.: Domain-specific configurations. For domains see Figure 4.3.

	domain			
feature	coarse	nest 1	nest 2	nest 3
grid mesh size	$0.25\,^\circ$	$0.125\,^\circ$	$0.0625\,^\circ$	$0.025\,^\circ$
horiz. grid cells	160×141	160×144	247×218	328×364
levels	40	40	40	50
lower left corner	$20^\circ W$,$20\,^\circ N$	$7\,^\circ W$,$33\,^\circ N$	$7\,^\circ W$,$34\,^\circ N$	$1\,^\circ W$,$36\,^\circ N$
time step	40 s	40 s	40 s	15 s
radiation update	1 h	1 h	1 h	15 min
cloud physics	1 mom	1 mom	1 mom	2 mom
spec. nudging	✓	-	-	-

Table 4.3.: Setups for scenarios. Further details of domain specific configurations are listed in Table 4.2.

short name	domain specific configuration	sea salt particles
CO	coarse	✓
N1	nest 1	✓
N2	nest 2	✓
REF ALL	nest 3	✓
REF WO	nest 3	-

In contrast to the domain-specific configuration of the coarse domain, the nested domains are configured on a rotated grid in which the North Pole is located at $170\,^\circ W$, $40\,^\circ N$. The configurations of the nested domains are not continuous but are split into three smaller time steps ("periods") of three days. The time step starting on November 2 is period 1, the time step starting on November 5 is period 2, and the last time step that starts on November 8 is period 3. Each nest configuration uses initial and boundary data from the next coarser domain size, which is updated every hour. No spectral nudging is applied to these configurations.

In the configuration for nest 3, the order for horizontal advection is increased from 3 to 5. Also, the time interval of the radiation calculations is reduced from 1 h to 15 min. The two-moment cloud micro-physics scheme in the configuration for nest 3 is used instead of the default bulk water-continuity scheme used in the other configurations (further described in section 3.2). Convection is parameterized by the Tiedtke-Scheme (Tiedtke, 1989), based on equilibrium closure of mass-flux with moisture convergence. In contrast to the other simulations, only the convection of shallow clouds is parameterized in the configuration of nest 3. Here, the grid mesh size is small enough to resolve deep convection.

Gaseous and aerosol emissions are taken into account in all domain-specific configurations. Emissions of mineral dust and sea salt particles are calculated based on the state of the atmosphere (see subsection 3.3.2). Other emissions of aerosols and gaseous compounds are taken from VEU (Kugler et al., 2013). All configurations neglect the direct feedback of aerosol particles on radiation because it is not of interest here.

The initial and boundary values for ART compounds (aerosol and gaseous concentrations) are taken from the next coarser domain, similarly to the meteorological fields. The exceptions are the initial ART fields of periods 2 and 3 in the nesting configuration. Here, the fields from the last time step of the period before are used.

Within the configuration of nest 3, a further distinction based on the treatment of aerosols is included. Besides the case, for which all technically available aerosols (see section 3.3) are included, a further configuration is added, in which all particles but sea salt are considered. The two scenarios resulting from the configurations of nest 3 are reference cases (REF ALL and REF WO, see Table 4.3) because these scenarios with the highest resolutions are of most interest.

The choice of the reference case is crucial. The entire removal of sea salt particles in the case REF WO gives a clear signal because the large emission and transport of sea salt particles due to the cyclonic wind field do not take place. Because other aerosols are represented in REF WO there is a consistent description of activation and micro-physical processes within the cloud as cloud nuclei are still present. A different reference case could have been chosen. For example, one could use a climatological background value of sea salt particles that is homogeneously distributed. However, the impact of sea salt particles would then be more difficult to see, because after the activation of the particles to cloud condensation nuclei these particles would then be similarly transported to the sea salt particles in the explicit description (REF ALL).

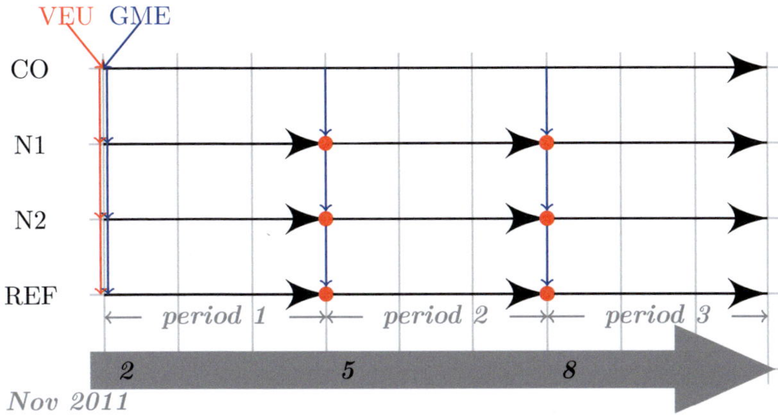

Figure 4.4.: Timeline for scenarios (overview in Table 4.3). The black arrows indicate the simulation time span. The blue arrows indicate the provision of initial data for meteorological fields. The red arrows indicate the provision of initial data for aerosol and gases. The filled red dots imply that initial data for aerosol and gases is taken from the period before.

4.3. Validation of Input Data

Data from the global model GME (Majewski et al., 2002) is used as the initial and boundary conditions of the meteorological variables. The model uses an icosahedral-hexagonal grid with 60 levels and an average grid spacing of approximately 20 km. Several weather services in Europe and developing countries use GME data for initial and boundary conditions in limited area models.

The input data is validated for two variables: sea surface temperature and low-level horizontal wind. The wind speed and direction demonstrates whether the Medicane, with its cyclonic structure and high wind speeds, is captured by the input data. The sea surface temperature is important for the genesis and development of the Medicane because most the energy results from the air–sea interaction.

To compare the sea surface temperature, the global daily dataset of the Jet Propulsion Laboratory Regional Ocean Modeling System (JPL

To compare the sea surface temperature, the global daily dataset of the Jet Propulsion Laboratory Regional Ocean Modeling System (JPL ROMS) group called Group for High-Resolution Sea Surface Temperature (GHRSST) is used. The dataset has a spatial resolution of 1 km×1 km.

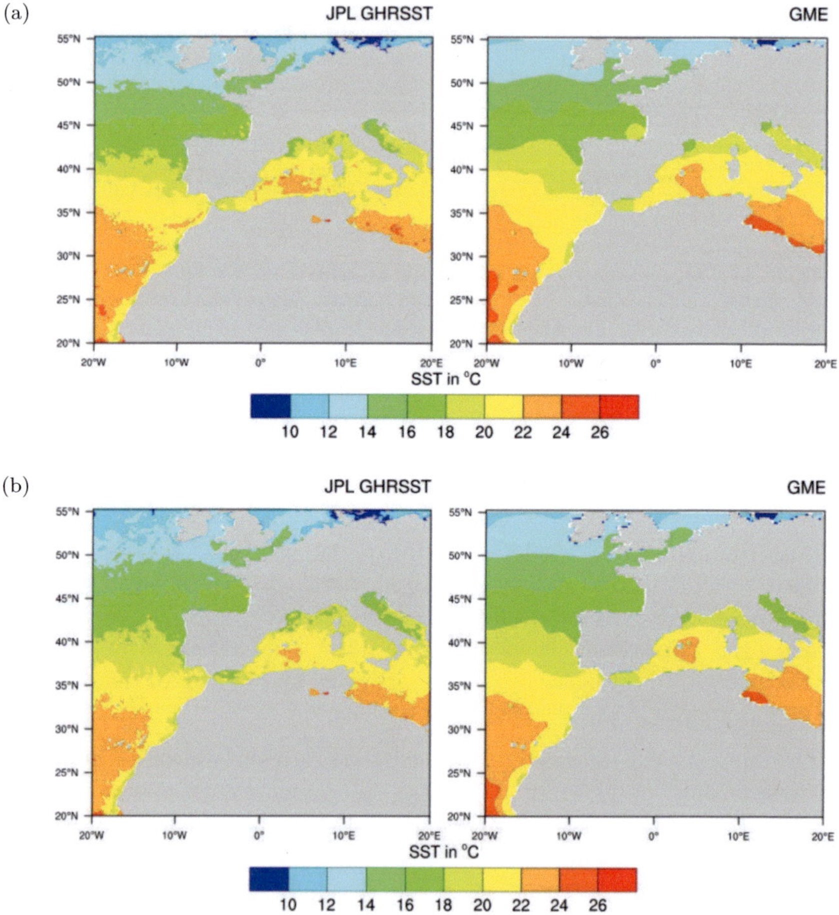

Figure 4.5.: Comparison of sea surface temperature. Left: Analyzed from JPL GHRSST. Right: daily mean of sea surface temperature in the GME model. (a) For November 3, 2011, and (b) for November 6, 2011.

The underlying data are measured by infrared sensors with resolutions of 1 to 2 km (e.g., Advanced Very High Resolution Radiometer (AVHRR), Moderate Resolution Imaging Spectroradiometer (MODIS), Advanced Along-Track Scanning Radiometer (AATSR)), geostationary satellites with a resolution of 6 km (GOES and METOP), and microwave sensors with a 25 km resolution (e.g., Advanced Microwave Scanning Radiometer Earth Observing System (AMSR-E), Tropical Rainfall Measuring Mission Microwave Imager (TMI)). In-situ point measurements by ships and buoys also contribute to the dataset. All of the available data are combined by a multi-scale two-dimensional variational blending algorithm with RMS errors of less than 1 ° C.

Figure 4.5 shows a comparison of the sea surface temperatures by JPL GHRSST and GME data. Both datasets are clipped to the coarse domain of the simulation (section 4.2). Analysis from the National Centers for Environmental Prediction (NCEP) are used for the sea surface temperature field in GME. In the GME case the data is interpolated to a latitude-longitude grid and calculated the daily mean of the 3 h fields. The sea surface temperature for the two days in Figure 4.5 and for the remaining days (not shown) are in good agreement. Of course, the higher resolution of the JPL GHRSST dataset allows for finer details, but the coarser structures are visible in both datasets.

The wind speed and wind direction near the surface of GME are compared with NOAA Blended Sea Winds data (Zhang et al., 2006) provided by the National Oceanic and Atmospheric Administration (NOAA) of the United States. The NOAA dataset contains 6-hourly wind vectors at 10 m height on a global 0.25 ° grid. These data have been available since July 9, 1987. The instruments are the Special Sensor Microwave Imager (SSMI), TMI, AMSR-E and Quick Scatterometer (QuickSCAT). Because of the low accuracy of the scatterometer-derived wind retrievals for wind speeds higher than 20 m/s (Ebuchi et al., 2002), the comparison is qualitative character rather than quantitative.

Figure 4.6 shows the NOAA Blended Sea Winds, clipped to the coarse domain and the horizontal wind vectors in the lowest model layer of GME. The wind vectors for both datasets are plotted for wind speeds higher than 8 m/s. Both points show a generally good agreement for the wind speed and wind direction. On November 4, 2011, at 18 UTC (Figure 4.6(a)), the

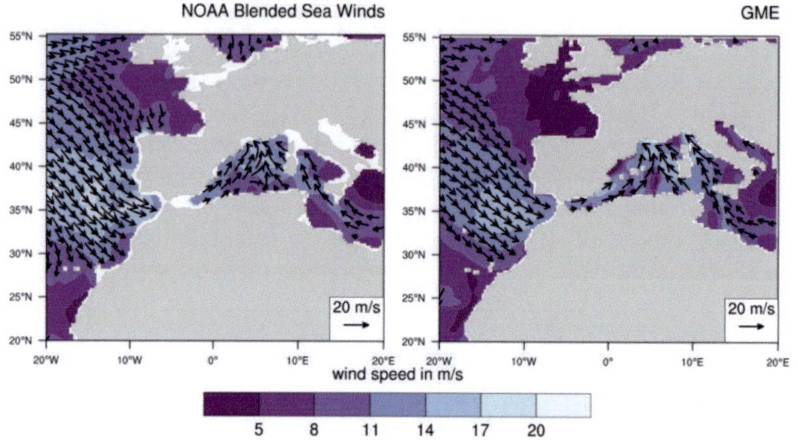

(a)

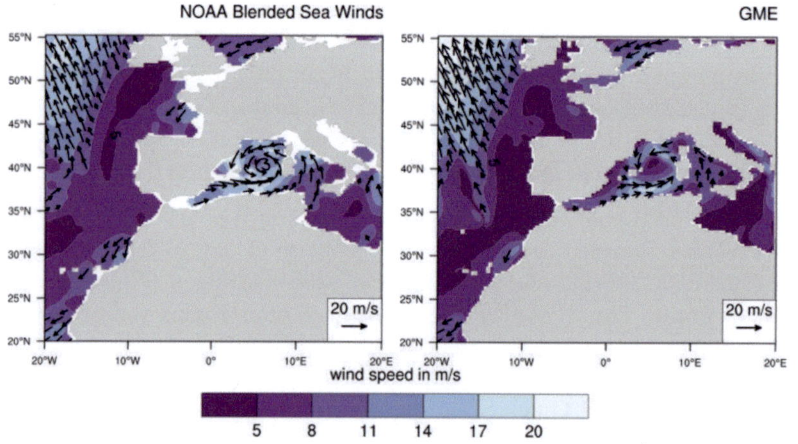

(b)

Figure 4.6.: Comparison of wind speed and wind direction over sea. Left: Analyzed wind in 10 m height from NOAA Blended Sea Winds. Right: Wind in GME model in approximately 10 m height. (a) For November 4, 2011, at 18 UTC and (b) for November 7, 2011, at 12 UTC.

GME wind speed was lower than the NOAA wind speed over the Celtic Sea and the Bay of Biscay. A lower wind speed is also visible on November 7, 2011, at 12 UTC (Figure 4.6(b)), in the North Atlantic west of Portugal and North Africa. At this time, the cyclonic Medicane is clearly visible between the Balearic Islands and Corsica in both datasets. Here, the NOAA Blended Sea Winds also indicate higher wind speeds. The core of the cyclone seems to be displaced northward in GME. Additionally, a strong wind speed gradient near the eye is visible.

In summary, the near-surface horizontal wind field and the sea surface temperature of GME agree well with the satellite data. This findings is optimal for the successful simulation of the synoptic situation. In the next section, a comparison with measured data verifies the simulation results of the Medicane.

4.4. Verification of Model Results

Because a real case is simulated, the simulation results must be validated. Since COSMO-ART is operated without data assimilation, the simulation must address the basic features of the atmospheric situation, i.e., the cyclone track, the eye, the axis-symmetric cyclonic cloud structure, the high wind speeds, and precipitation.

The following paragraphs show a comparison of the result of the simulation of nest 3 with sea salt particles (REF ALL, see Table 4.3) with observations. Because a forecast is beyond the scope of this study, the parameterizations within the model are not adjusted.

4.4.1. Comparison of Wind Field with Blended Observations

The horizontal wind near the surface is compared to the so-called NOAA Blended Sea Winds data (Zhang et al., 2006). This data was also used for validating the input data (section 4.3 and Figure 4.6).

Figure 4.7 shows that in scenario REF ALL, the wind speed is slightly underestimated compared with NOAA winds on November 8 at 12 UTC.

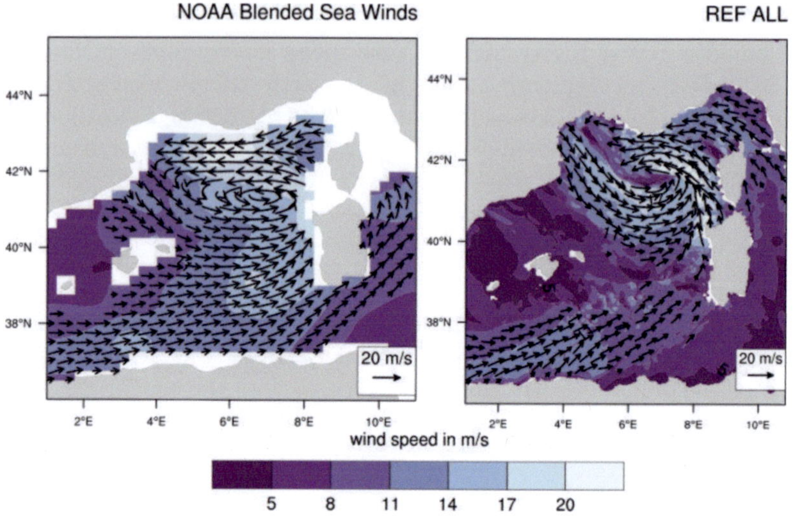

Figure 4.7.: Comparison of wind speed and wind direction over sea. Left: Analyzed wind in 10 m height from NOAA Blended Sea Winds. Right: Wind in scenario REF ALL in approximately 10 m height for November 8, 2011, at 12 UTC.

However, the agreement is satisfying. Additionally, the location of the eye of the Medicane is consistent between REF ALL and the satellite data.

4.4.2. Comparison of 3h Precipitation with Station Data

Data from SYNOP stations is used to verify the 3 h cumulative precipitation; buoy and ship measurements are not included. The direct comparison is visualized in Figure 4.8.

On November 6 at 6 UTC (Figure 4.8, top left) the simulation is similar to the measurements. For November 7 at 6 UTC (top right), the simulation shows precipitation over the sea, where there are no measurements. Precipitation was also simulated over Corsica, but the corresponding measurement sites reported no rain. On November 8 at 6 UTC (bottom), precipitation was simulated over the French coast, where most of the stations show precipitation amounts within the same range. Because of the lack of measurements over the seat, where much precipitation occurs,

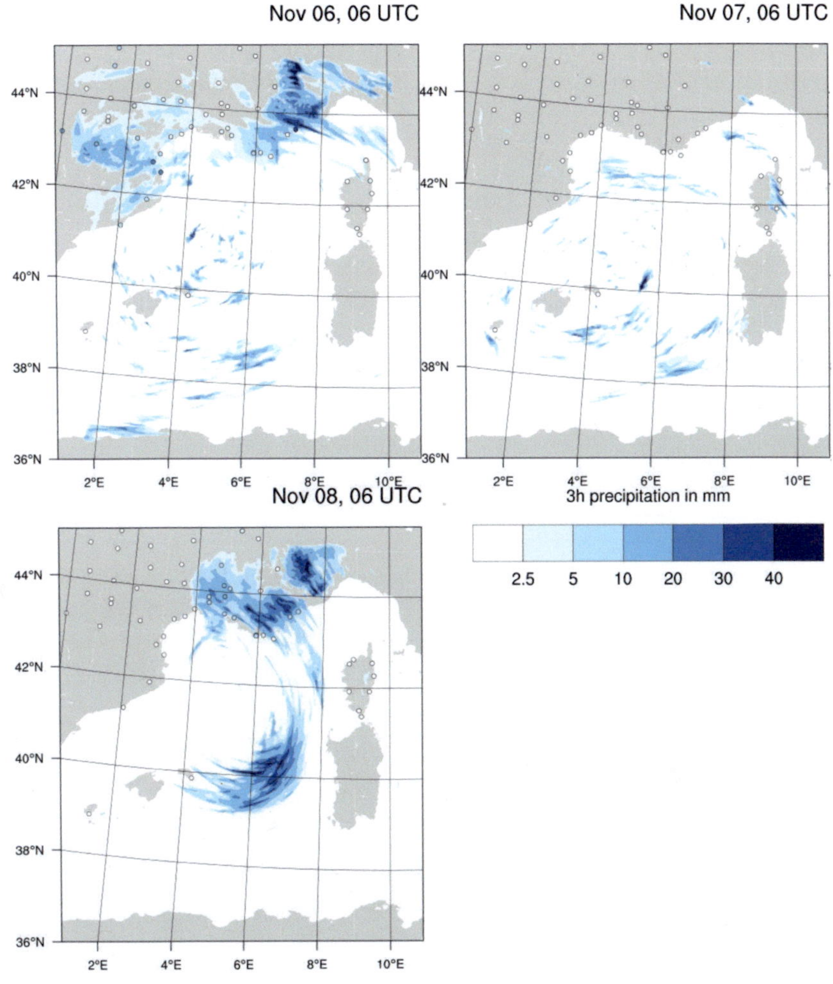

Figure 4.8.: Comparison of 3 h cumulative precipitation in mm of scenario with sea salt particles (contour) and SYNOP station data (circles) for three different days at 6 UTC: November 6 (top left), 7 (top right), and 8 (bottom).

it is difficult to conclude the degree of consistency between the simulation and observations. Over land, where the measurement network is available but unevenly distributed, the simulation sufficiently agrees with the observations.

4.4.3. Comparison of Brightness Temperature of Clouds with Satellite Images

Synthetic satellite images (see subsection 3.1.2) from the Spinning Enhanced Visible and Infrared Imager (SEVIRI) instrument of the Meteosat Second Generation (MSG), or Meteosat 9, (Eyre, 1991; Saunders et al., 1999) are used for cloud comparisons with real data from Meteosat 9(Schmetz et al., 2002). This geostationary satellite is located over the zero-degree meridian and therefore covers the entire simulation domain. The satellite produces images every 15 minutes at a spatial resolution of 3 km.

For the comparison, data of the IR channel between 9.8 µm and 11.8 µm (central wave-length of 10.8 µm) is used. This value is within the infrared atmospheric window; therefore, it measures the thermal emission of clouds. Based on Planck's law, the radiation temperature is calculated.

The synthetic satellite image for November 7 at 21 UTC is shown in Figure 4.9 (left). The real satellite image is displayed on the right for approximately the same domain. The color scheme is the same for both images. A cloud-free eye is visible in both images, and a large cloud band spans the entire domain from northeast to southwest. In the simulation, the radiation temperature is higher than the satellite measurement. Thus, clouds are simulated at lower height than observed. The positions of the eye and cloud bands agree quite well.

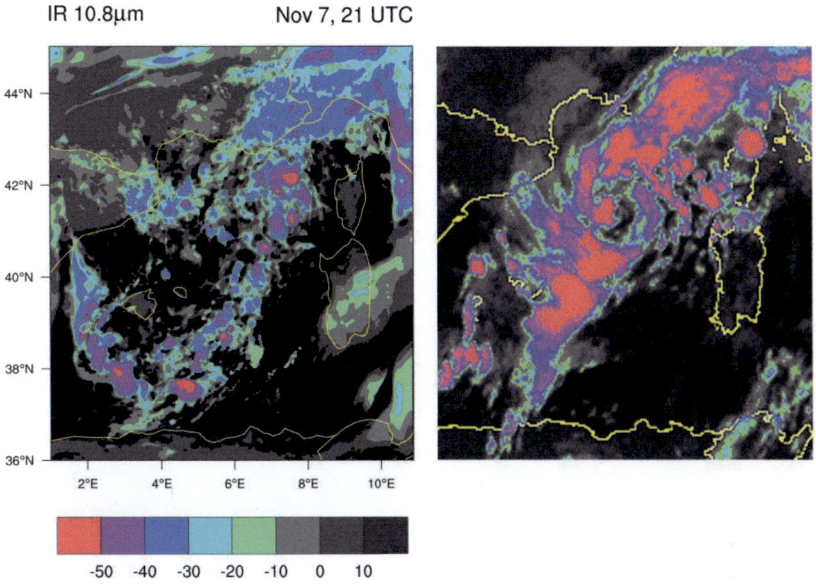

Figure 4.9.: Left: Brightness temperature in °C of synthetic satellite image on November 7, 21 UTC of SEVIRI channel 10.8 μm on MSG. Color legend is valid for both images. The original is taken from Mathieu Barbery (2011).

4.4.4. Comparison of Effective Radii of Cloud Droplets with Satellite Data

The effective radius r_{eff} is defined as ratio of the third to the second moments of a droplet or ice particle size distribution $n(r)$ given by

$$r_{eff} = \frac{\int_0^\infty r^3 n(r) \mathrm{d}r}{\int_0^\infty r^2 n(r) \mathrm{d}r}, \tag{4.1}$$

where r is the radius of the droplet or ice particle. The effective radius is important because it determines many radiative properties of clouds (e.g. albedo).

The effective radii from the scenario with sea salt particles are compared to data obtained from the Moderate Resolution Imaging Spectroradiometer (MODIS, King et al., 1992) on the AQUA satellite (Parkinson, 2003).

MODIS has 36 spectral bands that range from visible to infrared. The cloud mask of MODIS determines the location of the cloud and thus the validity area of all other cloud properties, the cloud-top properties and the cloud optical and micro-physical properties. Cloud-top properties are used in the algorithms of cloud optical and micro-physical properties. The default band combination for the effective radius retrieval is the 2.1 µm (water vapor-absorbing) and the 0.86 µm (non-absorbing) band.

Figure 4.10(b) shows the effective radii for cloud droplets on November 5, 2011. Data from four flights over the area are used to calculate the radii, which can be seen by AQUA in Figure 4.10(a).

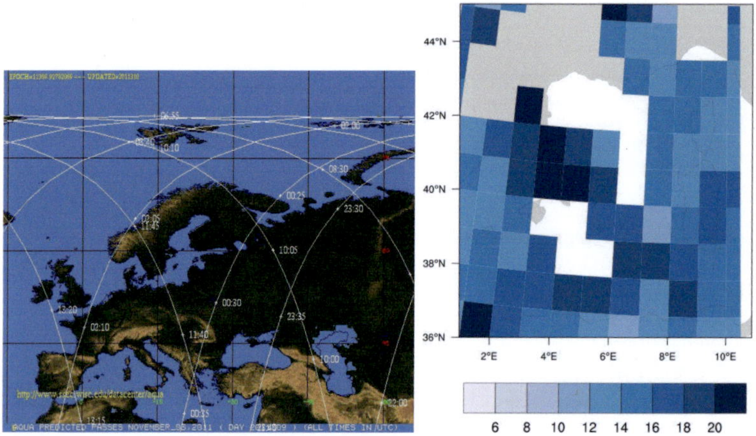

(a) (b)

Figure 4.10.: (a) Track and overflight times of AQUA satellite on November 5, 2011, taken from University of Wisconsin-Madison (2011). (b) Effective radii in µm for liquid particles, measured by MODIS on AQUA on November 5, 2011.

There are some major problems with MODIS cloud optical and micro-physical properties, as listed in the following paragraphs. For liquid water, the effective radii 30 µm and larger are considered failed retrievals. Additionally, the values of approximately 20 µm are uncertain according to Platnick et al. (2003). Using a different band (1.6 µm or 3.7 µm instead of the default 2.1 µm) to determine the effective radii results in differences of ±5 µm for both liquid water and ice particles over large areas in the case study of Platnick et al. (2003). In the Medicane case the cloud top temperature is mostly below -40 °C and thus the cloud top is frozen. This can be an additional source of uncertainty in the effective radii product of MODIS.

A uniform cloud with one vertical constant bulk effective radius is assumed by satellite retrievals. However, this assumption is not realistic. Therefore, to which height the satellite retrieval has to be assigned is unclear, which complicates comparisons with the model data. The uniform plane-parallel cloud assumption also leads to uncertainties in comparisons with clouds with their complex three-dimensional structure. The influence is large (for example, see Zhang et al., 2012)). The mean over an area of 5 km by 5 km can be highly biased compared to the domain average (up to 8 % in King et al., 2013). Drizzling clouds are also problematic in calculating the effective radii (King et al., 2013). Overall, MODIS overestimates the effective radii by typically 13 % (King et al., 2013).

Compared with simulation data (Figure 4.11), the effective radii have the same order of magnitude. For the simulation data, profiles of the vertical distribution of effective radii were used to emphasize the fact that there is no uniform distribution of the effective radii.

4.5. General Characteristics of the Simulated Medicane

This section presents the general characteristics of the simulated Medicane. Figure 4.12 shows the horizontal distribution of sea salt particles and wind vectors at the lowest model layer (approximately 10 m) for various domain-specific configurations (Table 4.2) on November 7, at 18 UTC. The wind vectors are plotted for wind speeds higher than 8 m/s. The cyclonic structure is visible in all domains; thus, the model is capable of simulating the Medicane in all setups. The highest number of sea salt particles corresponds to the locations of the highest wind speed in all

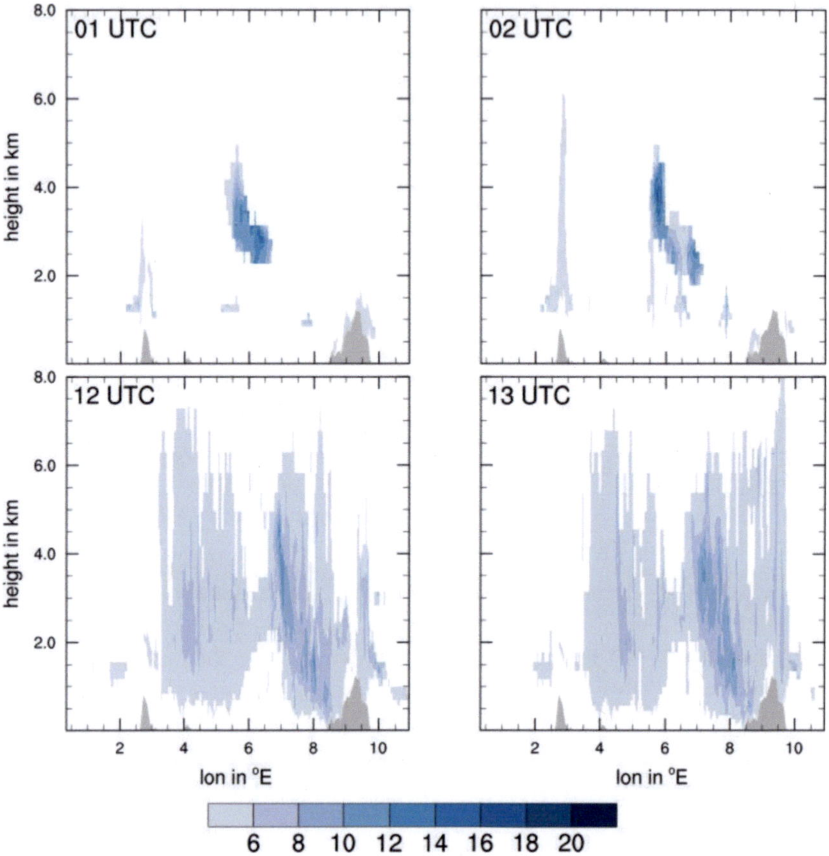

Figure 4.11.: Simulated effective radii in µm for liquid particles for scenario with sea salt particles (REF ALL) on November 5, 2011, at four different times corresponding roughly to the overflight times of AQUA (Figure 4.10(a)). The values in the latitudinally are averaged from approximately 39 to 40° N. The grey shading denotes the orography.

domains. Therefore, the Medicane produces and maintains its own sea salt field.

Figure 4.13 shows the temporal change in the composition of aerosols as the area average of the lowest model layer of scenario REF ALL (see Table 4.3). Over the entire simulation time, sea salt particles are the most

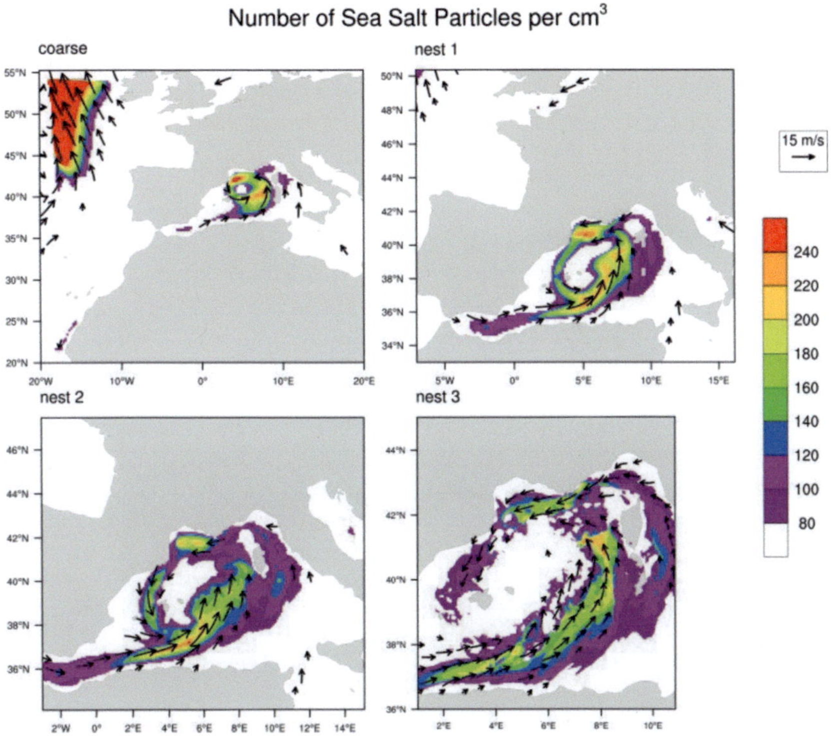

Figure 4.12.: Horizontal distribution of sea salt particles and wind vectors for at least 8 m/s in the lowest model layer on November 7, at 18 UTC for different simulation domains.

abundant aerosol, with a relative contribution of 80 to 60 %. The absolute mass concentration of sea salt is approximately 30 µg/m³. Unspecified anthropogenic aerosols follow, with a maximum contribution of 20 % (10 µg/m³). This aerosol class consists of anthropogenic aerosols in the coarse mode, particulate matter with a maximum diameter of 2.5 µm (PM2.5) and elementary carbon. Nitrate contributes approximately 10 % of aerosols. Primary organic compounds, soot, and ammonium have small contributions, whereas the remaining components nearly contribute to the total mass concentration of aerosols.

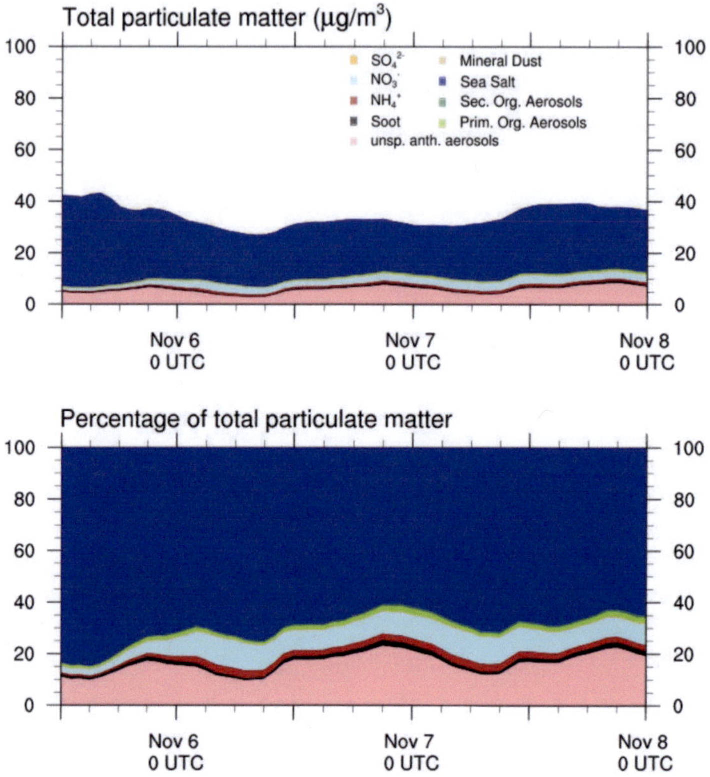

Figure 4.13.: Total particulate matter over time, starting on November 5, 12 UTC.

The vertical structure of the Medicane is characterized by a slight asymmetry according to the north-south component of the wind in a longitude-height cross section through the eye of the Medicane on November 7, at 18 UTC (Figure 4.14). The position of the eye is 4.75 °E at the surface. East of the eye, the v-component is southerly. West of the eye, it is northerly. The result is a cyclonic structure of the wind field around the eye. Near the surface, the highest wind speeds occur at approximately 8 km from the center (0.1 °). A second maximum is visible for a distance of approximately 40 km. At heights of approximately 5 km a westerly tilt of the eye region is found. Around 3 km, the v-component of the wind is smallest.

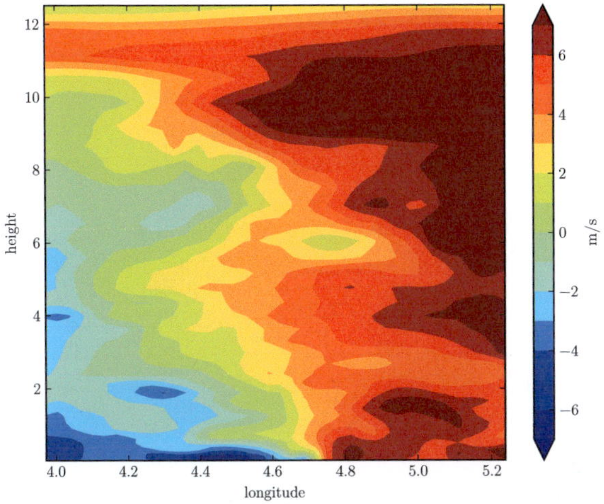

Figure 4.14.: Longitude-Height cross section of v-component of wind through eye of Medicane on November 7, at 18 UTC.

Displacements in different scenarios make it difficult to show direct differences at a given time in variables that exist only in specific areas. For example, cloud droplet numbers or effective radii of cloud droplets only exist within clouds. Large differences occur in areas where only one of the scenarios has a non-zero value possibly leading to wrong conclusions. Choosing averaged values over longer time periods in order to maximize those specific areas of a variable could cancel out important features. The next section displays the direct difference in precipitation between REF ALL and REF WO and discusses these and further complications.

4.6. Uncertainties Within Numerical Models

To study the impact of sea salt particles on precipitation, the differences between REF ALL and REF WO is considered as first step. The difference in the cumulative precipitation over 48 h (Figure 4.15(a)) corresponds to the cyclonic structure of the Medicane. The largest differences occur along the French coast and over the Mediterranean Sea (between 3 to 8 °E, 38 to 42 °N). Regions with anomalous precipitation can be equally distinguished.

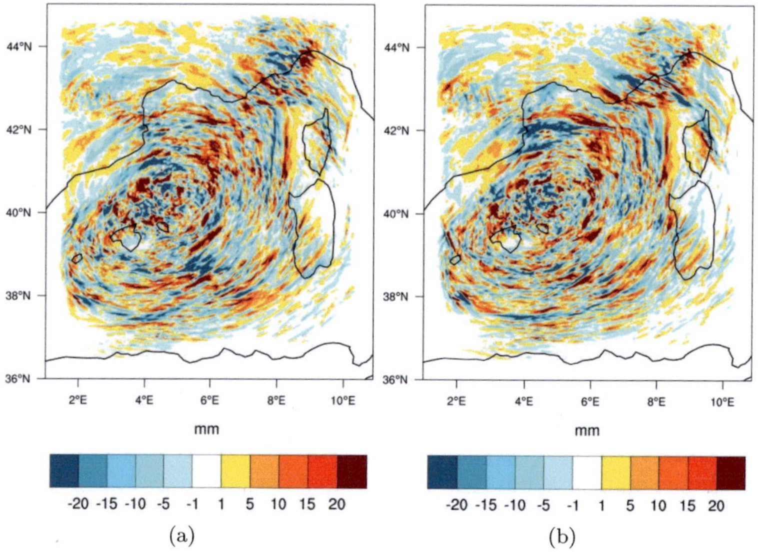

Figure 4.15.: Difference in 48 h cumulative precipitation for (a) scenarios REF ALL and REF WO and (b) REF ALL and REF ALL with an initial random perturbation of the temperature field.

To account for superimposed effects, a repetition of the simulation for scenario REF ALL with one small change were performed: a random perturbation is added to the time derivative of the temperature in the first time step. A more detailed description of this perturbation follows in subsection 5.1.2. The difference in the 48 h cumulative precipitation for REF ALL and the perturbed REF ALL is plotted in Figure 4.15(b). Compared with the plot to the left, both scenarios take sea salt particles into account. The difference is only caused by the perturbation. As a result of the solution of the Navier-Stokes equations, the perturbation propagates at the speed of sound (see subsection 5.1.2). The differences look very similar to those on the plot to the left. The cyclonic structure of the Medicane is visible and regions with anomalous precipitation of similar magnitude compared to Figure 4.15(a) are visible. The largest differences occur within the Medicane and along the French coast, although the positions and intensities for specific objects differ. The size range in which differences occur are the same for both cases.

Consequently, the differences in Figure 4.15(a) cannot be easily attributed to the impact of sea salt particles alone. There are also contributions from other factors that contribute to the impact of sea salt particles on precipitation. These factors have different sources that can be categorized into two classes. Epistemic uncertainties have their sources in missing knowledge and model limitations (Grimmett and Stirzaker, 2001). Thus, they can theoretically be reduced by model improvements. These uncertainties are due to the spatial and temporal discretization and the numerical solver: truncation error, linear damping, numerical dispersion and diffusion, aliasing, and non-linear instabilities. The second class contains aleatory uncertainties. They are conceptually not comprehensible and cannot be reduced (Grimmett and Stirzaker, 2001). The chaotic system linked to the non-linear equations of dynamics (Navier-Stokes equation) belongs to this category. It is known for a long time (Lorenz, 1963) that small changes in the initial conditions propagate fast in a non-linear system resulting in a diverging outcome. Thus, based on the individual system, there is a limitation in the prediction time. This is true even if the system is deterministic, which means that the system is fully described by the initial conditions and has no random elements.

This thesis deals with the treatment of aleatory uncertainties and suggests a method to separate the contribution of the non-linear dynamics from the impact of sea salt particles. As can be seen in Figure 4.15, the difference of only two aerosol realizations is not adequate for studying aerosol effects. There is a need for a more complex and sophisticated methodology to discriminate these impacts. The following chapter describes such a methodology.

5. A New Methodology for Separating Aerosol Effects on Precipitation

Assessing the impact of aerosols on clouds and precipitation is difficult (Stevens and Feingold, 2009). It is known for a long time (Lorenz, 1963) that due to the non-linearity of the Navier-Stokes equations slightly modified initial conditions can lead to completely different outcomes (see also Wang et al., 2012), even when considering aerosols. Thus, ensemble of models generated by perturbing initial conditions were proposed to determine the impact of aerosols (Morrison, 2012). As shown previously, differences in many variables occur over nearly the entire model domain, even in regions where only low concentrations of aerosol particles are modeled. The system is more sensitive to perturbations in some regions, regardless of the source of perturbation (see Figure 4.15). In these regions, it is very difficult to distinguish the effects of aerosols.

Classical grid-cell-based scores are often inappropriate for determining the differences in variables that have strong gradients on small spatial scales, such as precipitation (Ahijevych et al., 2009). Thus, other methods are needed. In the following, two methods for developing an ensemble and two scores that are used both alone and combined to identify the impact of sea salt particles on precipitation are discussed..

5.1. Ensemble Methods

For numerical weather prediction (NWP), a single realization of a model prediction is insufficient; instead, several setups with slight differences (e.g., in initial conditions or physical parameterizations) are nowadays used simultaneously, i.e., a model ensemble. The forecast quality is often determined by the "spread" of a variable, which shows the evolutions of this variable over time for the ensemble members. As long as the members exhibit similar values for a variable, the spread is small and a reliable forecast can be made. If the spread becomes larger, the members evolve differently and the forecasts are less reliable. There are many ways to create ensembles, and the proper method depends on the individual purpose.

The entire ensemble process accounts for the uncertainties in the model and input. Uncertainties due to the non-linear chaotic character of the dynamics are accounted for by the ensemble members with perturbations in the initial fields. To overcome uncertainties due to the driving model, ensemble members are given perturbations in the boundary conditions. By disturbing the model physics, uncertainties in the parameterizations are accounted for. Uncertainties inherent in the models are accounted for by a multi-model ensemble. As suggested by Morrison (2012), this study uses ensembles with initial-field perturbations.

5.1.1. COSMO-LEPS as Driver for COSMO-ART

One method for creating an ensemble is to change the initial and boundary data driving the model. This method aims to account for the non-linear dynamics and the uncertainties in the driving model. In a non-linear system, small changes in the initial conditions can have a strong impact on the system's response (Lorenz, 1963). Because this effect overlays the effect of aerosol, it has to be separated. For the reference scenarios of REF ALL and REF WO (section 4.2), the driver is scenario N2. This driver is replaced with an ensemble called "COSMO-LEPS" (Montani et al., 2011), which has 16 members. This creates a new ensemble called "LD" (**LEPS** **D**riven) by performing simulations based on the domain-specific configuration for nest 3 (Table 4.2). The resulting ensemble has 32 members, 16 of which are nested into one driving COSMO-LEPS member that considers sea salt particles; the remaining 16 members neglect these particles. This section describes the development of COSMO-LEPS.

COSMO is the model system used to perform the simulations (section 3.1). The **L**imited-area **E**nsemble **P**rediction **S**ystem (**LEPS**) was developed by the HydroMeteoClimate Regional Service of the Environmental Agency of Emilia-Romagna (ARPA-SIMC) in Bologna, Italy. The aim of the system is to help forecasters with alerts for events with a low deterministic predictability. Through nesting in a global-ensemble system, the advantages of the probabilistic approach are coupled to high-resolution details of the local area model. Therefore, it is possible to identify the occurrence of localized and intense weather events. COSMO-LEPS has been run operationally since November 2002 (Marsigli et al., 2005). The system works as a dynamical-downscaling global probabilistic model that reduces overall computational costs, as described later. Using this technique, the number of members decreases (from 102 to 16) without losing much information stored in the neglected (non-representative) members (Molteni et al., 2001).

COSMO-LEPS is driven by the global ensemble system of ECMWF called the Ensemble Prediction System (EPS), which has 51 members that each run with slightly different initial conditions. One model provides the control forecast, but it has no superior role compared with the other members in this study. The different initial conditions are disturbed through the singular vector technique (Buizza et al., 2008). The horizontal grid size is 80 km by 80 km, with 40 vertical layers and an integration time of 132 h.

For one complete COSMO-LEPS integration starting at a given day at 12 UTC, two sequential ECMWF-EPS runs are analyzed. The first run starts on a given day at 0 UTC and the second runs starts at 12 UTC. The result is an ensemble with 102 members.

The members are arranged into 16 clusters. These clusters constitute differing numbers of ensembles. To assign each member to a cluster, a complete-linkage method, which is a hierarchical cluster analysis, is used (Wilks, 1995). To determine which member belongs to which cluster, the variables must be discriminated. The variables include a combination of the two horizontal wind components, the geopotential height, and the specific humidity at 500, 700 and 850 hPa at the forecast steps of 96 and 120 h (relative to the latest ensemble run, which is 12 UTC of the given day).

The procedure to generate the discriminating variables is as follows:

- For each member m and for each of the variables $\varphi(x,y)$ at each level z and forecast step t, the mean over the clustering area A (30 °N-60 °N, 10 °W-30 °E) is calculated

$$\overline{\varphi^m_{z,t}} = \frac{1}{N_A}\sum_A \varphi^m_{z,t}(x,y), \tag{5.1}$$

  where N_A is the number of grid cells in the clustering area. The difference at each grid point value is divided by the standard deviation σ^m_φ. This results a dimensionless field $\phi^m_{\varphi;z,t}(x,y)$:

$$\phi^m_{\varphi;z,t}(x,y) = \frac{\varphi^m_{z,t}(x,y) - \overline{\varphi^m_{z,t}}}{\sigma^m_\varphi} \tag{5.2}$$

$$= \frac{\varphi^m_{z,t}(x,y) - \overline{\varphi^m_{z,t}}}{\sqrt{\frac{1}{N_A-1}\sum_A \left(\varphi^m_{z,t}(x,y) - \overline{\varphi^m_{z,t}}\right)^2}}. \tag{5.3}$$

- The quadratic distances Φ^{jk} among members j and k are computed for each of the fields at all levels and time steps

$$\Phi^{jk}(x,y) = \sum_\varphi \sum_z \sum_t \left(\phi^j_{\varphi;z,t}(x,y) - \phi^k_{\varphi;z,t}(x,y)\right)^2. \tag{5.4}$$

  Again, the mean over the clustering area is calculated as

$$\overline{\Phi^{jk}} = \frac{1}{N_A}\sum_A \Phi^{jk}(x,y), \tag{5.5}$$

  which results in the discriminating variables $\overline{\Phi^{jk}}$.

The cluster analysis is performed on the basis of these discriminating variables (the number of discriminating variables is 5151).

In the next step, a representative member is selected within each of the 16 clusters. The representative member of a cluster is the member that satisfies the following criteria:

- The ratio between the mean distance of the member and the mean distance of the other members of the same cluster is minimized.
- The ratio between the mean distance of the member and the mean distance of the members of the other clusters is maximized.

Hence, 16 representative members are selected, one in each cluster. These members identify important features of the associated cluster.

These representative members serve as initial and boundary conditions for the COSMO integration. One COSMO run is performed for each representative member to build the COSMO-LEPS. Integration starts on a specific day at 12 UTC. The horizontal grid size is 10 km by 10 km, and there are 40 vertical layers. The time step is 60 s, and the integration time is 60 h. The model results are output every three hours.

To account for model uncertainties, each of the 16 COSMO integrations randomly chooses a convection scheme, either Kain-Fritsch (Kain and Fritsche, 1993) or Tiedtke (Tiedtke, 1989).

The resulting COSMO-LEPS members are used as initial and boundary conditions for the meteorological fields in the LD ensemble. The initial aerosol and gas fields are taken from the reference scenarios (Table 4.3). Within LD, there are no reference cases. In contrast to what is stated in section 4.2 regarding the domain-specific configuration of nest 3, the LS ensemble updates the boundary fields every 3 h, starting at 12 UTC, and lasts only 60 h because the input from COSMO-LEPS is stored at this frequency. Figure 5.1 presents an overview of the simulation time span for LD and the handling of the initial data for both the COSMO and ART fields.

5.1.2. Initial Temperature Perturbations in COSMO-ART

Another method for creating an ensemble is to add a perturbation to a specific variable at a given time step. With this method, the ensemble members behave identically up to the point in time at which the perturbation is applied. Afterward, the behaviors may differ. Soon after the perturbation, the members are very similar, and differences are not easily seen. As the forecast time proceeds, differences increase, depending on the individual case, resulting in completely different states of the atmosphere. Because the perturbation is added at the first time step, this method can be seen as a perturbation of the initial conditions and therefore can also account for the non-linear dynamics.

This method of ensemble generation is applied on the reference scenarios described in Table 4.3 by adding a variation in the time derivative of the temperature at the first time step as described below. This ensemble has the name "ITP" for **i**nitial **t**emperature **p**erturbation ensemble. Three members with and without sea salt particles are created. Along with the reference cases, the entire random number ensemble has eight members.

In COSMO, the time integration is split into two modes to be numerical efficiency. As the fast sound waves are part of the solution in the compressible model atmosphere, they limit the time step of explicit time integration schemes. Thus, the prognostic equations are separated into terms which directly account for sound waves and slowly varying modes of motion. The time integration is separately performed for the modes. The time step for the slowly varying modes is subdivided into smaller time steps which are used for the fast modes.

An additional term is added to the time derivative of the temperature for slowly varying modes of motion s_T^{old}, which do not include variations that propagate with sound waves:

$$s_T = s_T^{\mathrm{old}} + \overline{s_T^{\mathrm{old}}} \cdot N_r. \tag{5.6}$$

$\overline{s_T^{\mathrm{old}}}$ is the mean value of the three-dimensional time derivative of the temperature at a given time. A random number perturbation is chosen to account for the uncertainty in the initial data. The three-dimensional random number field N_r is seeded with the system clock of the super computer. The elements are truncated-Gaussian distributed in the size range of -2 to 2. The perturbation is performed once in the simulation after the first time step for each grid cell.

Due to the time-splitting approach, the perturbation which becomes part of the overall model state contributes to the fast varying modes in the following. Thus, the perturbation also propagates with the speed of sound waves although it was originally implemented in the slowly varying modes only.

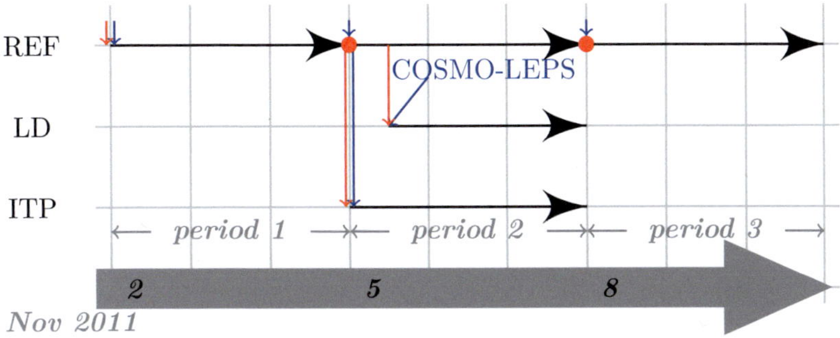

Figure 5.1.: Timeline for ensembles (see Figure 4.4 for timeline for scenarios). The black arrows indicate the simulation time span. The blue arrows indicate the provision of initial data for meteorological fields. The red arrows indicate the provision of initial data for aerosol and gases. The filled red dots imply, that initial data for aerosol and gases is taken from the period before.

This method of generating an ensemble could easily be applied to variables other than temperature. Because of the ideal gas law, changes in temperature affect pressure and density. Additionally, the relative humidity and stratification depend on the temperature. Thus, temperature perturbations propagate very quickly into many aspects of the model system (Rieger et al., 2014) and are sufficient in studying the system behavior.

Note that, because of the random seed, "pairs" of included and excluded sea salt particles cannot be compared directly; the perturbation is different for each member. In Figure 5.1, an overview of the simulation time for both ensembles, ITP and LD, including the handling of the initial data, is given.

5.2. Method of Analysis

Ensembles are often analyzed by comparing the area averages of a variable for each member. Temperature is often chosen for this comparison. I am interested in the changes due to sea salt particles, which are assumed to be very localized. Additionally, the expected impact does not occur on only one direction, but it can lead to an increase in a variable's value at one location and to a decrease at another location. Thus, taking an area average could cancel out interesting features.

A further problem emerges for classical statistical scores, for example, the χ^2 test of independence. This test is used to prove whether the independence of a frequency distribution of a target variable from two scenarios is statistically significant on a given level. Independence in the probability theory means that the occurrence of one event does not affect the probability of another. In general, such tests assume a distribution function of the tested variable or a specific distribution function, such as the normal distribution (i.e., student's t-test). For example, precipitation is not normally distributed.

Classical scores in meteorology (or 2-dimensional fields) are grid-cell based; thus, these scores state comparisons of corresponding grid cells. No information on temporal or spatial shifts is provided. Considering precipitation, in which shifts occur due to the impact of aerosols, these scores suffer from the so-called "double penalty". As an example, imagine a feature that is predicted well in terms of intensity and timing but at the wrong location compared with the observations. Classical scores assign the feature to both a "missed event" and a "false alarm". This results in an unfairly low forecast score. Classical grid-point-based scores are thus unsuitable to state the quality of a precipitation forecast.

5.2.1. Displacement and Amplitude Score (DAS)

There are four types of scores that comply with the need for better scores to analyze precipitation. Scores types are tested in the Spatial Verification Methods Intercomparison Project (ICP, Ahijevych et al., 2009):

- **Fuzzy** or **neighborhood** techniques require features that are close in space, time or intensity. The score quantifies the matching of features in terms of closeness.
- **Feature** or **object-oriented** techniques identify features and compare their properties.

- **Scale-decomposition** scores separate single physical features by applying spatial filters (e.g., Fourier or wavelet). Therefore, different scales can be addressed explicitly.
- **Field** verification is performed by comparing (precipitation) fields with the help of optical flow algorithms. There is no need to identify separate features.

The score used in this work is called the "**D**isplacement and **A**mplitude **S**core" (**DAS**). DAS belongs to the class of field verifications by optical flow algorithms. Keil and Craig (2007) developed the score and implemented minor changes in 2009 (Keil and Craig, 2009)). DAS quantifies how well features are predicted in terms of position and intensity. A two-dimensional vector field is computed to deform (or "morphs") one image to match another image. This process displaces the features of the first image. The magnitude of the vectors provides a measurement for the displacement, whereas the difference between the morphed and second image describes the difference in the amplitude of the precipitation. An advantage of this technique is that single features belonging to both fields do not need to be identified.

The score was originally developed to compare numerical weather prediction with satellite measurements. I use the score in a different way because I am interested in differences induced by changes in aerosol scenarios. Thus, as input data for the calculation of DAS, the horizontally distributed precipitation field of two simulations is used: $P1(x, y)$ and $P2(x, y)$. Furthermore, a value for the "maximum search distance" $D_{\max}$ must be given implicitly by another variable, the subsampling factor F_{DAS}. This variable is the diameter of the search environment for the matching algorithm. Within this search distance, features are assumed to be the same but displaced in space. Features that are separated over a large distance are considered unrelated. Therefore, the maximum search distance must be chosen individually for the synoptic situation and the process of interest. Additionally, a threshold can be used to remove background values. As is true for all scores, the result is influenced by and depends on the chosen conditions.

The image does not undergo deformation in one step. Instead, the two images are first coarsened, meaning that a given number of pixels (here, the horizontal grid spacing) are aggregated to form a new, coarser pixel element. Afterward, an initial vector field for the coarse fields is calculated using a pyramidal matching algorithm (Zinner et al., 2008). The vector field is applied followed by a new grain that is finer than the first grain. This chain is repeated until the full resolution of the image is reached.

The graining process depends on the subsampling factor F_{DAS}. The lowest resolution (topmost pyramid level, the first graining) averages $2^{F_{\mathrm{DAS}}}$ pixels from the original image as one pixel element of the new coarse picture. The maximum search distance $D_{\max}$ is given as the dependency of the subsampling factor F_{DAS} by

$$D_{\max} = \sqrt{2} \cdot 2^{F_{\mathrm{DAS}}+2}. \tag{5.7}$$

The unit of $D_{\max}$ in Equation 5.7 is pixels of the original image. To obtain the maximum search distance, one must multiply $D_{\max}$ by the horizontal grid spacing.

The displacement vector for each pixel is computed. Within the range of ± 2 pixels, a translation in all directions is possible. A displacement that provides the minimum squared difference in a local region is chosen. To favor pixel elements that are closer to the original pixel, the extent of the local region is defined by the convolution of a two-dimensional Gaussian kernel with a standard deviation of $\frac{5}{3}$. The center is located on the original pixel element in the center. The vector field is then applied to the original image resulting in a temporary image. In the next step, the temporary image is coarsened by averaging $2^{F_{\mathrm{DAS}}-1}$ pixels from the original image into one pixel of the new image (which is the next pyramid level). A new vector field is computed and applied. The steps are repeated until the full resolution of the original image is reached. The sum of the individual vector fields is used to generate the final morphed image. The magnitude of the displacement vectors provides the displacement field $\mathrm{DIS}_{P1}(x, y)$. The root-mean-square difference of the final morphed image and the compared image provides the amplitude field $\mathrm{AMP}_{P1}(x, y)$. During the graining process and the vector field application, the mass of the precipitation is not completely conserved but rather marginally reduced.

A simple example of the process of coarsening, the calculation and the application of the displacement field is depicted in Figure 5.2. The precipitation field of simulation 1, $P1$, is shown in blue. Green is the precipitation

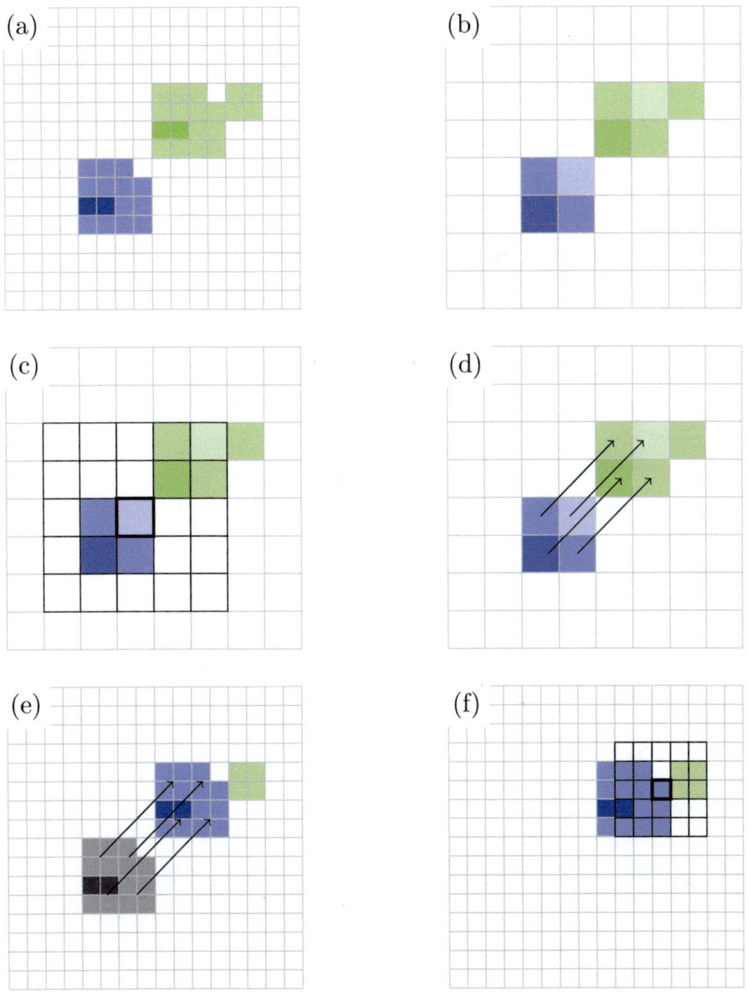

Figure 5.2.: In blue is the precipitation field $P1$ and in green $P2$ in (a) the original resolution and (b) coarse grained with $F_{\mathrm{DAS}} = 2$. (c) and (f) show the search distance for one specific pixel. The calculated displacement field is shown in (d) and applied on $P1$ in the original resolution in (e).

field of simulation 2, $P2$, visualized in the same grid space. In (a), the precipitation fields of the two simulations are displayed in the original space (grey grid). Coarsening with a subsampling factor of 2 is performed. Thus, 2 by 2 pixels from the original space form a new pixel in the coarsest space (the lowest resolution), as shown in (b). Afterward, for each coarse pixel in $P1$ within the space of ± 2 pixels, a corresponding pixel in $P2$ is searched. The search distance for one specific pixel (thick black border) is represented in (c) as thin black borders. A vector field links the corresponding pixels of $P1$ with $P2$, as shown in (d). This vector field is applied to $P1$ at the original resolution (grey shading in (e)), which results in a temporary, intermediate precipitation field (blue shading in (e)). Afterward, the process of coarsening and the calculation of the displacement vector field are repeated until the original resolution is reached. In our example, these criteria is reached at this step. The corresponding pixels are searched within the original resolution for the last time, as shown in (f) for one specific pixel. The displacement field is calculated and applied. The difference of the resulting morphed image of $P1$ and $P2$ is transferred to the amplitude field AMP_{P1}. The sum of the magnitude of the displacement fields for each resolution states the final displacement field DIS_{P1}.

The two fields are zero where no features occur in the original picture. Non-zero values in $\text{DIS}_{P1}(x, y)$ and $\text{AMP}_{P1}(x, y)$ only appear in locations where precipitation occurred. A non-zero value of $\text{DIS}_{P1}(x, y)$ means that a feature was matched within the search diameter in the two images but was spatially displaced. A value of zero means either a perfect match or no feature match. A missed feature, when a precipitation pixel occurred in only one of the two simulations, results in a large value of $\text{AMP}_{P1}(x, y)$ at the location.

In contrast, how well the features of simulation 2 match the features of simulation 1 must be verified. Similarly, the fields $\text{DIS}_{P2}(x, y)$ and $\text{AMP}_{P2}(x, y)$ are calculated. Here a zero value in the displacement field and a large value in the amplitude field indicates a false alarm. That is again, when a precipitation object occurred in only one of the simulations, in this case in the other one compared to the missed feature discussed in the paragraph above.

The displacement score over the entire verification area A in the space of simulation 1 is given by

$$\overline{\mathrm{DIS}_{P1}} = \frac{1}{N_{P1}} \sum_{A} \mathrm{DIS}_{P1}(x, y). \tag{5.8}$$

N_{P1} is the number of grid cells whit a non-zero value. By normalizing according to N_{P1}, the score is insensitive to the size of the verification area A or to the area of non-events. The displacement score in the space of simulation 2 is similarly defined. An amplitude score over the entire domain is obtained by calculating the root-mean-square average of the values of AMP_{P1}

$$\overline{\mathrm{AMP}_{P1}} = \sqrt{\frac{1}{N_{P1}} \sum_{A} \mathrm{AMP}_{P1}^2(x, y)}. \tag{5.9}$$

$\overline{\mathrm{AMP}_{P2}}$ is calculated respectively. The total displacement score is calculated by

$$\mathrm{DIS}' = \frac{1}{N_{P1} + N_{P2}} \left(N_{P1}\overline{\mathrm{DIS}_{P1}} + N_{P2}\overline{\mathrm{DIS}_{P2}}\right) \tag{5.10}$$

and the total amplitude score is calculated by

$$\mathrm{AMP}' = \frac{1}{N_{P1} + N_{P2}} \left(N_{P1}\overline{\mathrm{AMP}_{P1}} + N_{P2}\overline{\mathrm{AMP}_{P2}}\right). \tag{5.11}$$

In a later step, these individual scores must be normalized before they form the final displacement-amplitude score. The displacement field is normalized with the maximum search distance $D_{\max}$ (Equation 5.7). For the amplitude score, a characteristic intensity I_0 that depends on the individual application is used. I_0 is the root-mean-square amplitude of the precipitation field of simulation 1, $P1(x, y)$,

$$I_0 = \sqrt{\frac{1}{N_{P1}} \sum_{A} P1^2(x, y)}. \tag{5.12}$$

The combination of the individual scores for the displacement (Equation 5.10) and amplitude (Equation 5.11) normalized by $D_{\max}$ (Equation 5.7) and I_0 (Equation 5.12) leads to the final displacement-amplitude score

$$\mathrm{DAS} = \frac{\mathrm{DIS}'}{D_{\max}} + \frac{\mathrm{AMP}'}{I_0} = \mathrm{DIS} + \mathrm{AMP}. \tag{5.13}$$

A perfect match of the two fields is given for DAS = 0. Values below zero are not possible. DAS ≈ 1 is typical, but there is no upper limit.

5.2.2. Factorial Method (FM)

Factorial approaches are used to study the impacts of various conditions on the outcome of an experiment. These conditions are represented by factors designed to reduce the number of experiment runs; the number of runs is large if the experiment design has many factors that contribute to the results. An experiment that depends on only two factors with two values each is more efficient than an experiment design that includes only one factor at a time. The possible values of a factor are called "levels" in statistics. If the factors interact in an experiment, then a factorial approach is inevitable. In this study, the individual model runs are the experiment of which is spoken in the following. In atmospheric sciences, there are two major factorial approaches.

A straightforward factorial approach is factor separation **f**actor **s**eparation (**FS**) (Stein and Alpert, 1993; Alpert and Sholokhman, 2011). Here, the contributions of factors are isolated from their interactions. This approach can identify the factor with the strongest impact on a specific outcome of an experiment.

The other factorial approach is called **F**actorial **M**ethods (**FM**) or factorial design (Montgomery, 2008). This approach is used in many scientific and engineering fields and in numerical modeling (Teller and Levin, 2008). Compared with FS, FM allows for a wide range of statistical analyses. Within FM, a complete replicate of an experiment means that all possible combinations of factors and levels are investigated.

FM can identify the contribution of a factor to the variability in an experimental result and can compare it to the contributions of other factors. Additionally, the theory of FM allows for multiple replicates of an experiment, as is required here (see details later in this chapter). Because FS lacks these features, it is not capable of representing my experiment design. Thus, I chose FM over FS.

The following summarizes information on FM by Montgomery (2008) that is relevant to this study.

A 2^2 factorial design is used, i.e., there are two factors at two levels. One complete replicate of this design requires $2 \times 2 = 2^2 = 4$ experiments. The result of an experiment that is performed for the i-th time is Ψ_i. It depends on two factors, sticking with the terminology of Montgomery (2008), they are written in capital Latin letters, A and B:

$$\Psi_i = fct(A, B) = \Psi_i(A, B). \tag{5.14}$$

If an effect has two levels, the values are called "low" and "high". The levels are represented by lowercase Latin letters (unlike the notation in Montgomery (2008)):

$$A \in a_1, a_2, \tag{5.15}$$
$$B \in b_1, b_2, \tag{5.16}$$

where a_1 and b_1 are the low level and a_2 and b_2 are the high level. Table 5.1 shows all possible combinations for an experiment.

Table 5.1.: Possible combinations in a 2^2 factorial design

Factor		
A	B	Label
a_1	b_1	$\Psi(a_1, b_1)$
a_2	b_1	$\Psi(a_2, b_1)$
a_1	b_2	$\Psi(a_1, b_2)$
a_2	b_2	$\Psi(a_2, b_2)$

The total of all N replicates is the sum over all replicates, e.g.

$$\Psi(a_1, b_1) = \sum_{i=1}^{N} \Psi_i(a_1, b_1). \tag{5.17}$$

$\Psi(a_1, b_2)$, $\Psi(a_2, b_1)$, and $\Psi(a_2, b_2)$ are calculated respectively. It is possible to display the treatment combinations graphically using a square (Figure 5.3).

The effect of a factor is denoted with the same uppercase Latin letter as the factor itself; however, in calligraphy the effect of factor A is denoted

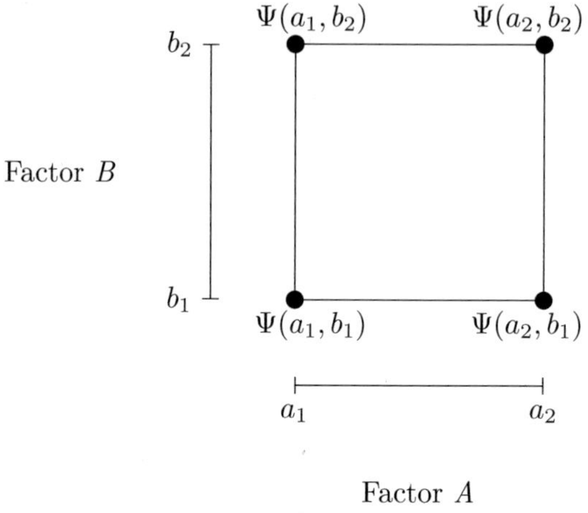

Figure 5.3.: Treatment combination in a 2^2 design.

by $\mathcal{A}$. The effect of the interaction of factor A and B is denoted by $\mathcal{AB}$. The effect of a factor refers to the change in the experiment result due to a change in the level of a factor, specifically, the difference in the average of the experiment outcome using this effect at the high level and the average of the outcome using this effect at the low level. For example, the effect of A at the high level is $(\Psi(a_2, b_1) + \Psi(a_2, b_2))/2N$ and the effect at the low level is $(\Psi(a_1, b_2) + \Psi(a_1, b_1))/2N$. The main effect of A is

$$\mathcal{A} = \frac{\Psi(a_2, b_1) + \Psi(a_2, b_2)}{2N} - \frac{\Psi(a_1, b_2) + \Psi(a_1, b_1)}{2N} \tag{5.18}$$

$$= \frac{1}{2N} \left(\Psi(a_2, b_1) + \Psi(a_2, b_2) - \Psi(a_1, b_2) - \Psi(a_1, b_1)\right). \tag{5.19}$$

Analogous to effect $\mathcal{A}$ (Equation 5.18), the main effect of factor B is calculated by

$$\mathcal{B} = \frac{1}{2N} \left(\Psi(a_2, b_2) + \Psi(a_1, b_2) - \Psi(a_2, b_1) - \Psi(a_1, b_1)\right). \tag{5.20}$$

The interaction effect $\mathcal{AB}$ is

$$\mathcal{AB} = \frac{1}{2N} \left(\Psi(a_2, b_2) + \Psi(a_1, b_1) - \Psi(a_2, b_1) - \Psi(a_1, b_2)\right). \tag{5.21}$$

The sign of an effect indicates its direction, i.e., whether an effect is responsible for an increase or decrease in the experiment results. Using the magnitude and direction of the effects $\mathcal{A}$, $\mathcal{B}$, and $\mathcal{AB}$ will help determine which variables are likely to be important. To confirm this first guess, **an**alysis **o**f **va**riance (ANOVA) can be used. A fundamental technique of ANOVA is the sum of squares, which are used to indicate the variance of each factor. The sum of squares is calculated by

$$SS_A = \frac{(\Psi(a_2,b_2) + \Psi(a_2,b_1) - \Psi(a_1,b_2) - \Psi(a_1,b_1))^2}{4N}, \tag{5.22}$$

$$SS_B = \frac{(\Psi(a_2,b_2) + \Psi(a_1,b_2) - \Psi(a_2,b_1) - \Psi(a_1,b_1))^2}{4N}, \text{ and} \tag{5.23}$$

$$SS_{AB} = \frac{(\Psi(a_2,b_2) + \Psi(a_1,b_1) - \Psi(a_2,b_1) - \Psi(a_1,b_2))^2}{4N}. \tag{5.24}$$

By dividing the sum of squares of a factor by the total sum of squares, one obtains the relative contribution of this factor to the total variability. For one replicate, $N = 1$, the total sum of squares, SS_T, is the sum of the sum of squares for all factors, Equation 5.22 to Equation 5.24

$$N = 1: \qquad SS_T = SS_A + SS_B + SS_{AB}. \tag{5.25}$$

With more than one replicate, the total sum of squares is calculated by

$$\begin{aligned} &N > 1: \\ &SS_T = \sum_{i=1}^{n} \left(\Psi_i(a_2,b_1)^2 + \Psi_i(a_1,b_2)^2 + \Psi_i(a_2,b_2)^2 + \Psi_i(a_1,b_1)^2\right) \\ &\qquad - \frac{(\Psi(a_2,b_1) + \Psi(a_2,b_2) + \Psi(a_1,b_2) + \Psi(a_1,b_1))^2}{4N}. \end{aligned} \tag{5.26}$$

In contrast to Equation 5.25, with more than one replicate it is possible to calculate the experimental error. The sum of squares due to error SS_E contributes to the total sum of squares SS_T in Equation 5.26, which can also be written as

$$N > 1: \qquad SS_T = SS_A + SS_B + SS_{AB} + SS_E. \tag{5.27}$$

The sum of squares due to errors are calculated as the last term by rearranging Equation 5.27,

$$N > 1: \qquad SS_E = SS_T - SS_A - SS_B - SS_{AB}. \tag{5.28}$$

With a single replicate it is not possible to estimate the error this simple.

In Appendix B, there is a simple example of the usage of FM.

FM is applied on precipitation fields of the ITP ensemble (subsection 5.1.2) to separate the impact of sea salt particles from the impact of the initial perturbation, which accounts for effects of the non-linearity of dynamics. Previously, the use of area-averaged variables were criticized because they may cancel out interesting features. However, area averaging is performed here by applying FM. Note that the application of FM is not done alone; it is combined with DAS, which has localized information.

By applying FM to the ITP ensemble, one obtains two factors: sea salt particles and the initial perturbation that represents the impacts of the non-linearity of the dynamics. The interpretation of the interaction of both factors is more complicated: the factors influence each other, and they are not independent over time. Because the contribution of the interaction is a result of both factors, it cannot be stated as an impact due to sea salt particles alone. Although it is clear that sea salt particles contribute to the variance described by this variable, the impact cannot be separated here.

The experiment is repeated three times using the three members of the ITP ensemble for each scenario with and without sea salt particles. The experiment components with no perturbations produce the same result for each replicate, assuming that there are no uncertainties in the super computer. Thus, the error is contributed by the perturbation only.

5.2.3. A New Approach for Coupling the Methods DAS and FM

By applying DAS, differences in precipitation can be determined and quantified if they are caused by spatial shifts or intensity differences.

If and to what extent these differences are linked to aerosols remains unknown.

FM quantifies the impact of sea salt particles on differences in the area-averaged precipitation and separates it from those impacts due to the non-linear dynamics. Spatially resolved information is lost within FM.

To compensate the disadvantages, I combine the methods. First, DAS is applied to account for the type of differences in the precipitation. Second, grid cells are filtered that show the feature of interest to an extent (e.g., an displacement of at least 8 km) and then FM is applied to those grid cells. The feature is either the $\mathrm{DIS}_{P1}(x, y)$ or the $\mathrm{AMP}_{P1}(x, y)$ field of DAS for a chosen minimum value.

For example, grid cells that show a precipitation displacement of at least 8 km are chosen. Then, the average of the precipitation value of these grid cells is taken to calculate the relative contribution of the factors by FM. With this application, one can state that aerosols somewhat contribute to the changes in precipitation due to spatial shifts of at least 8 km.

The combination of DAS and FM permits, for the first time, a systematically study of the impact of aerosols on precipitation by separating the various effects and by determining the types of changes. The following questions can be addressed using this technique: Do aerosols lead to shifts in precipitation? How large are the shifts? On what time scales can an impact of aerosols on precipitation using a specific simulation be seen?

6. Advanced Model Based Case Study of Medicane "Rolf"

Increasing numbers of aerosols, which act as cloud condensation nuclei (CCN), lead to small cloud droplets because of the competition for water vapor. However, more cloud droplets can form. These droplets do not grow as efficiently as large droplets because collision–coalescence declines. Consequently, the droplets need more time to reach the size of raindrops, at which point they leave the cloud. This process may lead to a spatial shift of the precipitation compared with a scenario with few aerosols and CCN.

Droplets that remain in a cloud for a long time have a greater probability of reaching high altitudes. If the droplets arrive at levels where heterogeneous nucleation occurs, then the cloud can develop more vigorously and build large solid hydrometeors at the expense of precipitation in the liquid phase. If the CCN is large ($r > 1\ \mu m$), then large cloud droplets form and rain may start at lower levels and/or at earlier points in time. Nevertheless, if rain is delayed or accelerated, then the total amount of cumulative precipitation does not need to change because a dynamical feedback may be induced.

To determine whether the differences in a precipitation pattern caused by sea salt particles are related to spatial shifts or differences in the amount of DAS of Keil and Craig (2009) (see subsection 5.2.1) is applied. Six small rectangular areas build areas of special interest as subdivisions of the simulation domain of nest 3: (1) the Balearic Islands, including an area around Majorca, Minorca, Ibiza, and Formentera; (2) Corsica; and (3) Sardinia, with the nearby sea; (4) an area over the Mediterranean Sea without any land; and coastal regions of mainland Europe; (5) an eastern area where the Medicane caused the most damage; and (6) a western area (Figure 6.1).

6.1. Analysis of the Reference Scenarios

How fast the signal of sea salt particles in precipitation propagates is of interest. Spatial shifts are directly linked to this signal speed and are quantified based on the DAS calculations as follows. Spatial shifts could appear linked to the horizontal wind velocity because clouds are transported at this speed. However, shifts could also be linked to the transition velocity of the Medicane. Based on the Tropical Bulletins by the NOAA Satellite and Information Service (Table 4.1), the maximum transition velocity is 6.7 m/s, which results in a displacement of 72.4 km in 3 h.

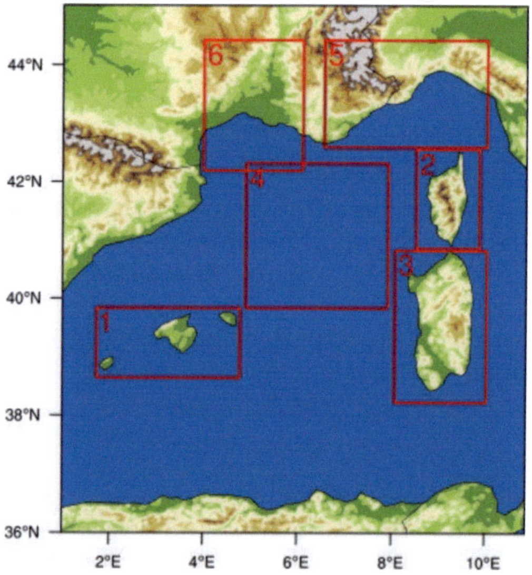

Figure 6.1.: Simulation area of domain specific configuration of nest 3 (Table 4.2). Surrounded by red rectangles are sub-domains, for which further investigations are done.

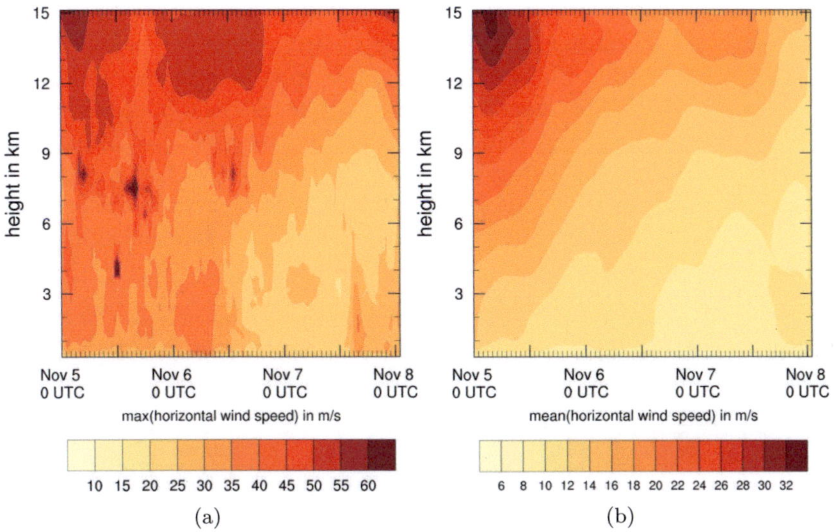

Figure 6.2.: Horizontal wind speed of scenario REF ALL over time, (a) maximum and (b) mean of each layer.

The order of magnitude of the shifts linked to the horizontal wind speed is estimated by extracting the maximum horizontal wind speed of each model layer over time for the scenario REF ALL (Table 4.3), which is plotted in Figure 6.2(a). A relatively high wind speed of 50 m/s, multiplied by a given time period, provides the upper value of possible displacements. Using this value for wind speed, displacements of up to 180 km in 1 h are possible. Assuming that this wind speed maintained over 3 h, displacements of up to 540 km could theoretically be reached within this time span. Averaging the wind speeds (Figure 6.2(b)) leads to smaller displacements: 20 m/s would lead to a displacement of 72 km in one hour or 216 km in 3 h. Taking the mean instead of the maximum wind speed produces more realistic displacements if the horizontal extension of the cloud covers several grid cells. The maximum wind speed reached by a single grid cell, but not by the surrounding grid cells, is inadequate to move the entire cloud. The surrounding grid cells must have similar wind speeds to shift the entirety of a cloud. Additionally, the values of the maximum wind speed do not last for 3 h, as seen in Figure 6.2(a). Thus, this value cannot be used to calculate displacements in 3 h. One has to keep in mind that shifts within the Medicane are not linear in one direction but are more circular. The linear distance, which is calculated with DAS, would then be smaller than a displacement along such a path. Typical lifetimes of clouds are between 30 min and 1 h; thus, different clouds may be identified when calculating displacements over 3 h. This outcome should not be a problem because the Medicane consists of many clouds, and identifying single clouds is not possible.

In summary, a signal speed linked to the transition velocity of the Medicane is approximately 6.7 m/s, a signal speed linked to average wind speeds results in a higher signal speed of approximately 20 m/s. Later in this chapter, those theoretical signal speeds are compared to the actual signal speed derived by the application of DAS.

6.1.1. Results of the Application of DAS

DAS is applied to the cumulative precipitation fields at different time spans of the reference scenarios with sea salt particles and of the scenario without sea salt particles (REF ALL and REF WO) for the entire nest 3 domain and the sub-domains. The score has two input parameters: a threshold for the precipitation and the so-called subsampling factor. Values below the threshold are handled as if there was no precipitation at all. The subsampling factor is coupled to the maximum search distance, which is the largest distance two grid cells can have to be considered the same precipitation object by the score. Grid cells with a large distance are assumed to be different, independent objects. The following section uses with the descriptive maximum search distance, although the abstract subsampling factor is the input parameter of DAS.

48 h Cumulative Precipitation

DAS is applied to the 48 h cumulative precipitation starting on November 6 at 00 UTC using the nest 3 setup. Values chosen for the threshold are of 25 mm, 50 mm, and 100 mm and for the subsampling factor of 1 and 2.

Inserting the subsampling factor in Equation 5.7 and multiplying by the horizontal grid spacing results in maximum search distances of 31.7 km and 63.4 km, respectively.

The cumulative precipitation field for the reference scenarios REF ALL and REF WO is plotted in Figure 6.3 (top left and top right). The figure refers to $P1$ (REF ALL) and $P2$ (REF WO) of subsection 5.2.1. The vector field, which is calculated to deform $P2$ to match $P1$, is overlaid onto $P2$ on the top right. The precipitation field after deformation by the vector field is shown on the bottom left (morphed $P2$). The difference between $P1$ and morphed $P2$ is used to calculate the amplitude field. Because the root-mean-square difference is used, the amplitude field, AMP_{P1} shows only positive values.

Table 6.1.: DAS, AMP, DIS, mean differences in intensity, AMP′, and mean displacements, DIS′, for nesting 3 domain on 48 h cumulative precipitation.

$D_{\max}$	th_{DAS}	DAS	AMP	DIS	AMP′	DIS′
31.7 km	25 mm	0.4557	0.2244	0.2313	12.24 mm	7.32 km
31.7 km	50 mm	0.3835	0.2067	0.1768	17.05 mm	5.60 km
31.7 km	100 mm	0.3264	0.2102	0.1162	27.86 mm	3.68 km
63.4 km	25 mm	0.3887	0.2122	0.1765	11.58 mm	11.18 km
63.4 km	50 mm	0.3102	0.1875	0.1227	15.47 mm	7.77 km
63.4 km	100 mm	0.2786	0.2113	0.0673	28.01 mm	4.26 km

As seen by the values of DAS in Table 6.1, the precipitation fields better agree for higher thresholds than for lower ones for both maximum search distances, $D_{\max} = 31.7$ km and $D_{\max} = 63.4$ km. A better agreement indicates that the precipitation field of the two scenarios differs only slightly; thus, the impact of sea salt particles is low. Differences due to displacements (DIS) decline for higher thresholds, whereas differences due to intensity (AMP) do not remarkably change.

The right two columns of Table 6.1 present the mean of the difference in the intensity (AMP′) and the mean of the displacement (DIS′) calculated using AMP and DIS when multiplied by norm factors (see subsection 5.2.1 and Equation 5.13), I_0 and $D_{\max}$. In this application of DAS, I_0 is 54.55 mm, 82.51 mm, and 132.54 mm for the thresholds of 25 mm, 50 mm, and 100 mm, respectively. For higher thresholds, the differences in the intensity increase while the displacements decrease. For a large maximum search distance, the displacements are slightly larger compared with a low maximum search distance, whereas the differences in intensity show no significant change.

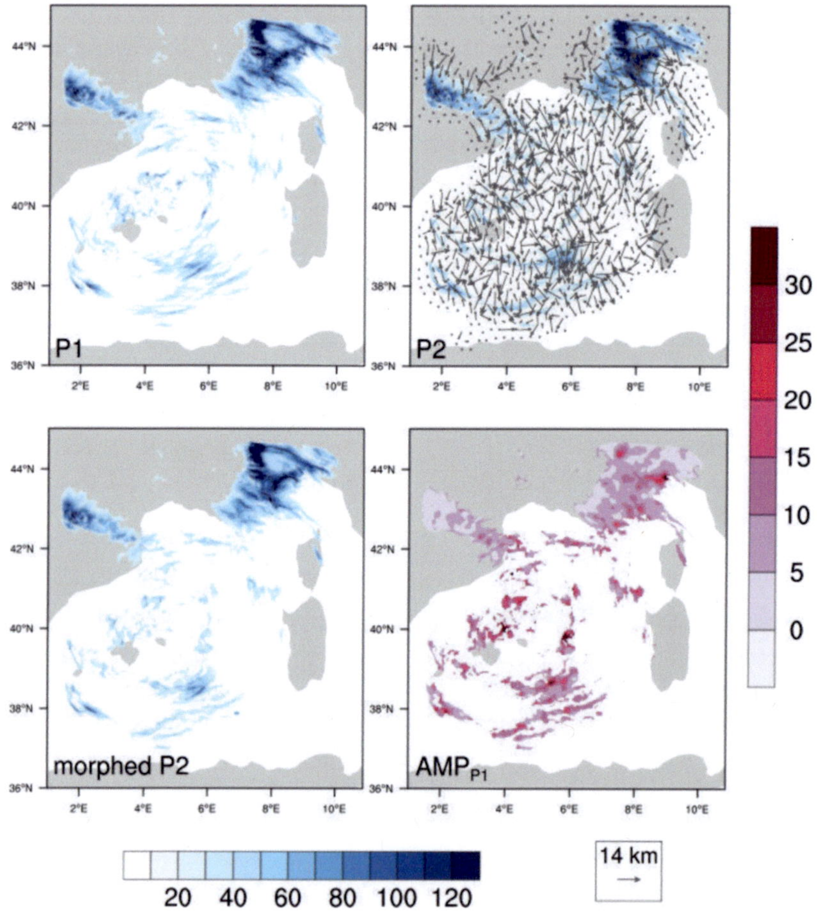

Figure 6.3.: 48 h cumulative precipitation in mm for the scenarios REF ALL ($P1$, top left) and REF WO ($P2$, top right) overlaid with the displacement vector field calculated with DAS. Bottom left shows the result of the application of the displacement field on REF WO (morphed $P2$). Bottom right is the resulting amplitude field in mm.

3 h Cumulative Precipitation

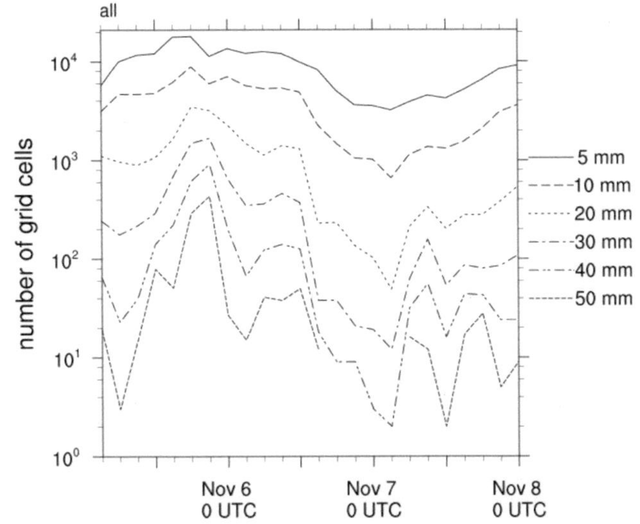

Figure 6.4.: Number of grid cells for several precipitation thresholds over time.

DAS is applied to the 3 h cumulative precipitation of the reference scenarios with the domain-specific configuration of nest 3 for period 2 (for specifications see section 4.2) starting on November 5. To consider the temporal change, all of the time intervals of period 2 are used, which results in 24 time intervals for the 3 h cumulative precipitation. For the setup of DAS, several thresholds and maximum search distances were assigned. Only thresholds that have an adequate number of grid cells for DAS are reasonable. Figure 6.4 shows the number of grid cells over time for thresholds of 5 mm, 10 mm, 20 mm, 30 mm, 40 mm, and 50 mm. The larger the threshold, the smaller the number of grid cells. Thresholds of 40 mm and 50 mm corresponds to less than 100 grid cells most of the time.

This number is too low for a reasonable application of DAS, which is why the following neglects those threshold values.

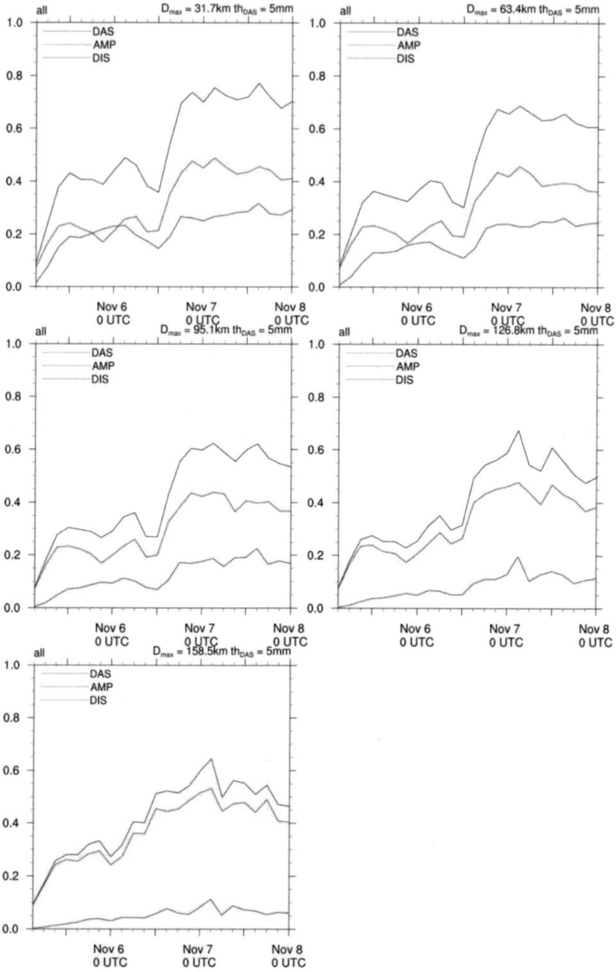

Figure 6.5.: DAS (black line), AMP (red), and DIS (blue) for the nesting 3 domain and a constant threshold of $th_{\text{DAS}} = 5\,\text{mm}$ for 3 h cumulative precipitation over time. Values used for threshold th_{DAS} and maximum search distance $D_{\max}$ are noted on the top right of each plot.

To determine the order of magnitude of displacements, a constant threshold of 5 mm is used and the maximum search distance is varied between 31.7 km and 158.5 km (which corresponds to subsampling factors of 1 to 5). The resulting values of DAS for the reference scenarios, REF ALL and REF WO, and the contributions by AMP and DIS are shown in Figure 6.5 as a time series. For all maximum search distances, DAS increases with time; therefore, the differences in precipitation increase over time. This finding is not astonishing because the initial states of the two scenarios are identical; afterwards, they develop freely. Thus, a longer simulation has a higher probability of developing unique conditions.

Increasing the maximum search distance decreases DAS because DIS also decreases as AMP slightly varies, with no connection to different maximum search distances. Differences in the precipitation field due to differences in the amplitude of the precipitation do not depend on the chosen maximum search distance. Nearly all displacements are covered by a maximum search distance of 31.7 km. If shifts have maximum values of 31.7 km, then no new attribution of corresponding precipitation objects occurs when a larger maximum search distance is chosen. DIS decreases with increasing maximum search distances (Equation 5.13). The factor used for normalizing AMP′ depends on the chosen threshold, which is kept constant in Figure 6.5. Because of this finding and because AMP does not increase or decrease for different maximum search distances, maximum displacements of 31.7 km are likely to occur.

Therefore, displacements do not occur on the order of magnitude estimated at the beginning of this chapter (several hundreds of kilometers within 3 h). Displacements do not depend on the horizontal wind speed or the transition velocity of the Medicane. Assuming maximum displacements of 31.7 km within 3 h results in a signal speed of approximately 3 m/s, which is much lower than that estimated. The following text sets the maximum search distance constant at 31.7 km and chooses thresholds of 5 mm, 10 mm, 20 mm, and 30 mm.

Figure 6.6 shows the calculated values of DAS for the reference scenarios, REF ALL and REF WO, over time. Additionally, contributions due to displacements (DIS) and differences in precipitation amount (AMP) are displayed. Here, different thresholds are chosen whereas the maximum search distance is constant at $D_{\max} = 31.7$ km. The value of DAS increases as the forecast time proceeds; this is also true when varying the maximum search distances, as shown previously.

The red line representing the differences in the precipitation amount has higher values than DIS nearly everywhere. Thus, changes in the amplitude are more important than changes in the displacement. In the beginning of the simulation, the displacement greatly contributes to differences in the 3 h cumulative precipitation. The distance between the two lines is very small, and the relative importance of both lines is nearly the same. At approximately 18 UTC on November 6, the distance increases and the relative contribution of DIS decreases.

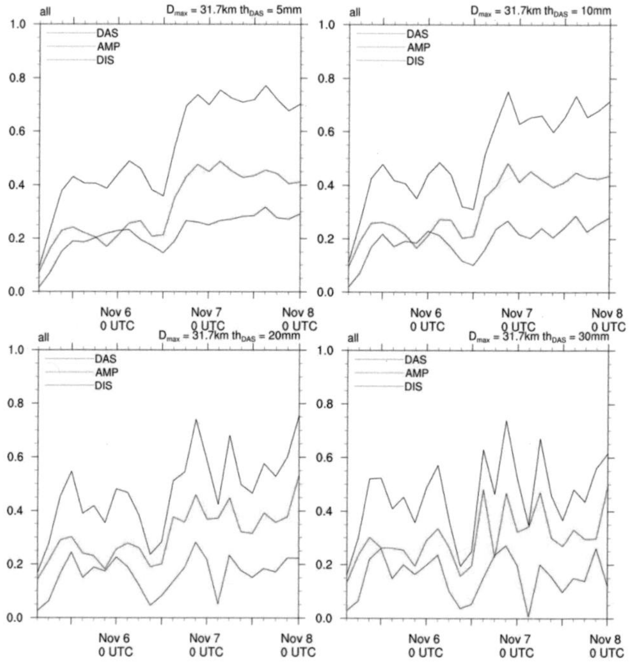

Figure 6.6.: DAS (black line), AMP (red), and DIS (blue) for the nesting 3 domain and a constant maximum search distance of $D_{\max} = 31.7$ km for 3 h cumulative precipitation over time. Values used for threshold th_{DAS} and maximum search distances $D_{\max}$ are noted on the top right of each plot.

From 12 UTC to 18 UTC on November 6, there is a large jump of DAS for all thresholds. This jump is also visible in DIS and AMP. For $th_{\mathrm{DAS}} = 5\,\mathrm{mm}$ and $th_{\mathrm{DAS}} = 10\,\mathrm{mm}$, the jump in AMP is larger than that in DIS, whereas for $th_{\mathrm{DAS}} = 20\,\mathrm{mm}$ and $th_{\mathrm{DAS}} = 30\,\mathrm{mm}$, the jump is larger in DIS than in AMP. The number of contributing grid cells in Figure 6.4 shows a clear decline of at least a factor of 2 (for a threshold of 5 mm) up to a factor of 10 (30 mm threshold). In addition, the total amount of precipitation was lower at 18 UTC than at 12 UTC (Figure 6.7 and Figure 6.8). Together, these two findings result in a smaller I_0 at 18 UTC than at 12 UTC. A small I_0, which is the normalization factor for AMP′ (Equation 5.12), leads to a large AMP. This finding could partly be responsible for the jump in DAS. The location of the precipitation objects could also account for the jump, as described in the following paragraph.

The 3 h cumulative precipitation at these points in time are visualized in Figure 6.7 and Figure 6.8. At 12 UTC, clearly there is more precipitation, particularly around the Alps and the Pyrenees. This orographical precipitation occurs in both scenarios at the same location. Thus, DIS has a low value here. At 18 UTC, there is no orographic precipitation around the Alps, and the intensity of precipitation is also reduced compared with six hours earlier. Thus, smaller and less intense precipitation objects over the Mediterranean Sea more greatly contribute to the calculation of DAS than comparable objects at 12 UTC. At 12 UTC, DAS is mainly determined by the high precipitation around the mountains, which is forcibly in good agreement in space and intensity; aerosols cannot lead to large shifts, which results in a small score. At 18 UTC, objects over the sea are more important. Here, aerosols can more easily influence precipitation because of the homogeneously flat environment.

The findings within this chapter for the simulation domain of nest 3 are conferrable to all sub-domains (Figure 6.1). The corresponding plots are in section C.1. One large difference is that DAS is zero at several points in time for the small domains. Precipitation does not occur everywhere during the entire simulation, so there are points when no rain occurs in specific sub-domains.

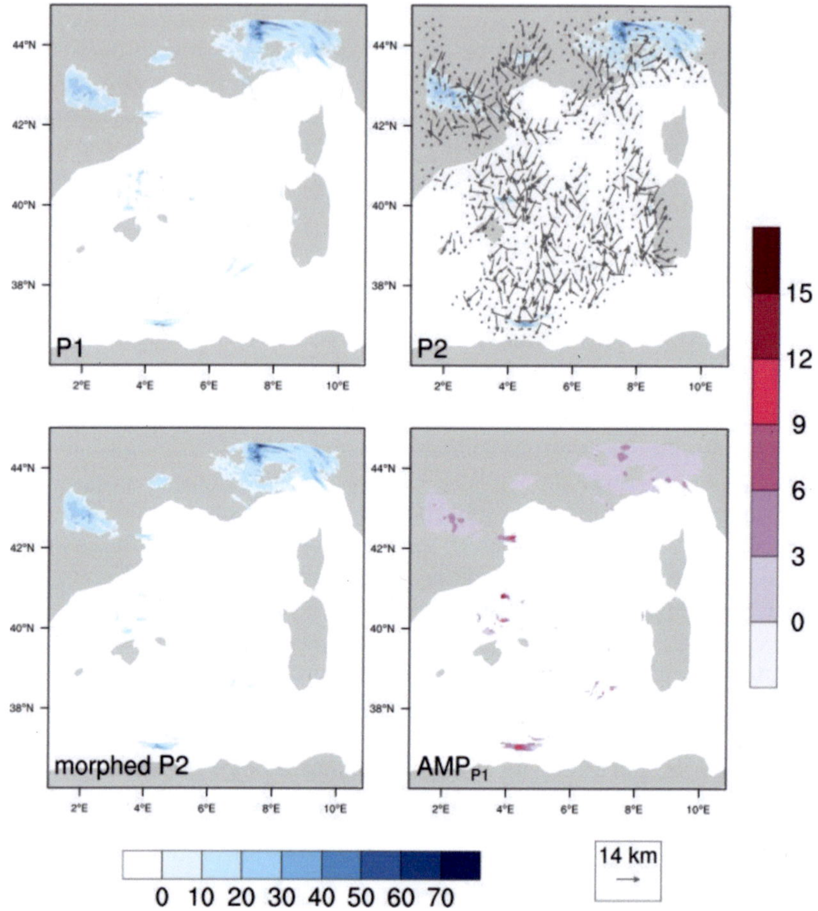

Figure 6.7.: 3 h cumulative precipitation in mm at 12 UTC on November 6 for the scenarios REF ALL (*P*1, top left) and REF WO (*P*2, top right) overlaid with the displacement vector field calculated with DAS. Bottom left shows the result of the application of the displacement field on REF WO (morphed *P*2). Bottom right is the resulting amplitude field in mm.

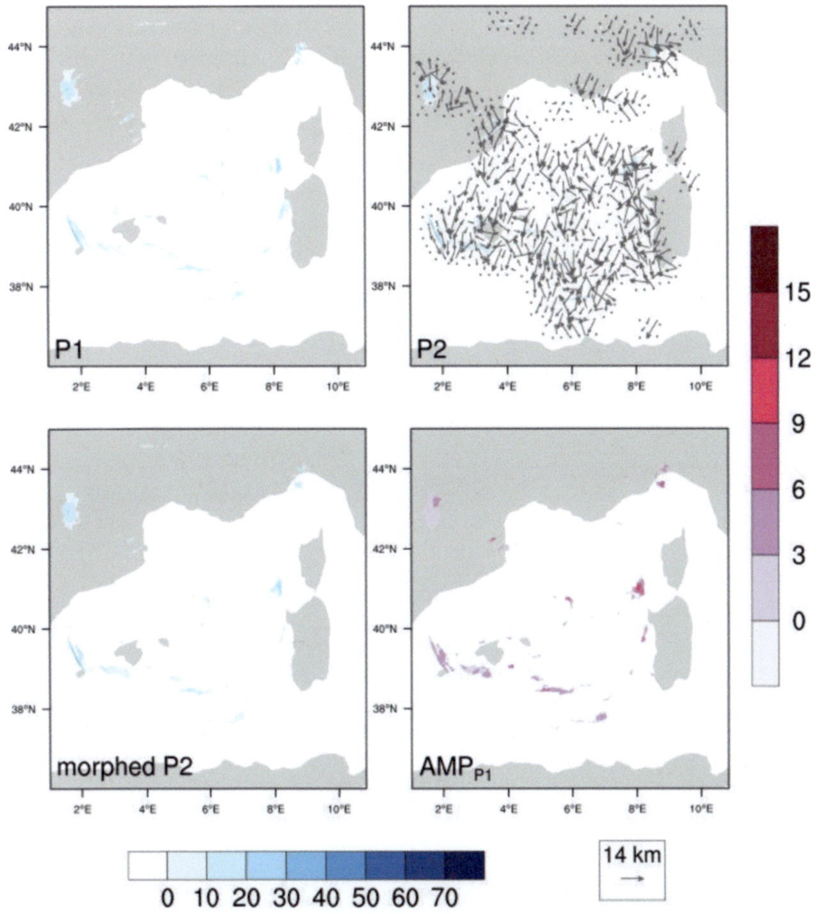

Figure 6.8.: Same as Figure 6.7 but at 18 UTC on November 6.

6.2. Analysis of the LEPS Driven Ensemble

An ensemble is created by changing the driver of the model. COSMO-LEPS (see also subsection 5.1.1) is the new driver for the scenarios with and without sea salt particles (the reference cases in section 4.2 in the nest 3 domain. Each of the 16 members are used as driver for both scenarios, to obtain an overall ensemble with 32 members. Because all of the original members are variable but equally important, no reference cases were included.

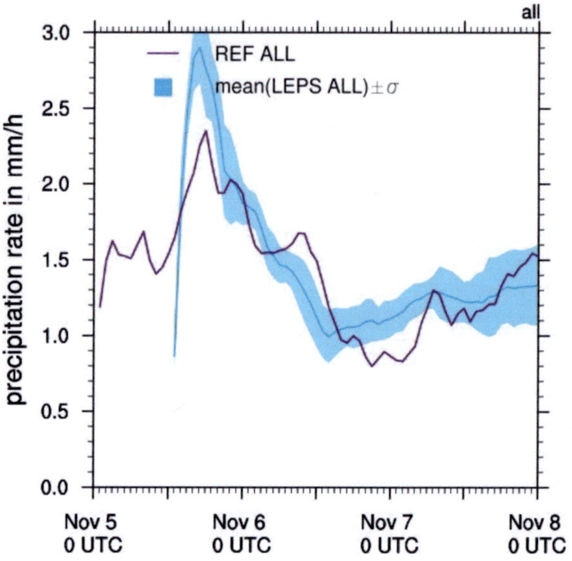

Figure 6.9.: Precipitation rates over time for scenarios taking sea salt particles into account covering nesting domain 3. Violet line shows scenario REF ALL. Blue line shows mean of LD members with sea salt particles. Blue contour covers the range of the standard deviation calculated with Equation 6.1.

Figure 6.9 shows the area-averaged precipitation rate for the reference scenario REF ALL and the mean of the (16) LD members with sea salt particles as solid lines. The area average is taken for the entire domain. The blue contour indicates the range of 2σ around the mean of the LD members (mean $\pm\sigma$), in which σ is the corrected sample standard deviation calculated with

$$\sigma = \sqrt{\frac{1}{N_m - 1} \sum_{i}^{N_m} (rr_i - \overline{rr})^2}, \tag{6.1}$$

where N_m is the number of ensemble members, rr_i is the precipitation rate of member i, and $\overline{rr}$ is the mean of the precipitation rates of all members. All grid cells that have a precipitation rate of less than 0.01 mm/h are neglected. The violet line indicates that the reference scenario REF ALL starts earlier than the LD ensemble (see timeline for ensembles in Figure 5.1).

A maximum value is visible at approximately 21 UTC on November 5 in LD and in REF ALL. Later, the precipitation rate decreases, with local maxima and minima at different points in time. Towards the end of the simulation, the precipitation rate increases again. The blue curve is smoother than the violet curve because of the additional mean of the LD members. The standard deviation, which is connected by the spread of the members, has only a small variation over time (0.2 to 0.3 mm/h). Although the values for both lines are rarely the same, the overall behaviors are very similar.

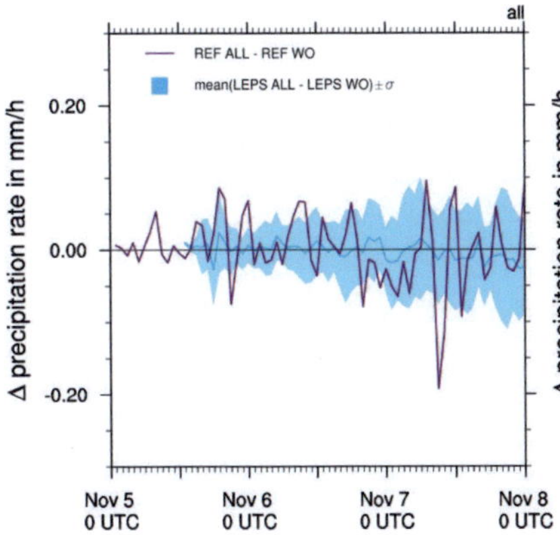

(a)

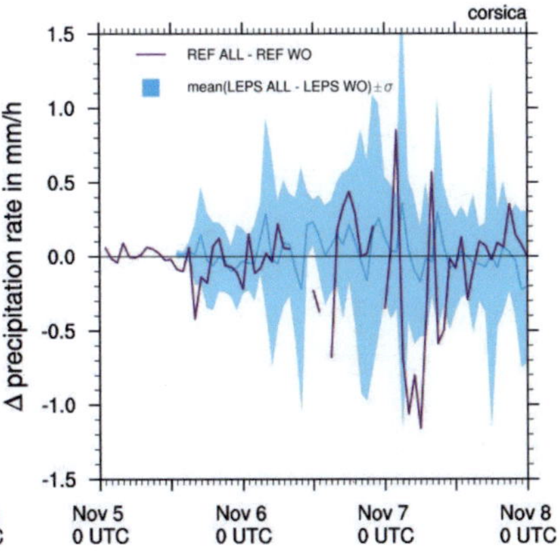

(b)

Figure 6.10.: Difference in precipitation rate over time of scenarios with and without sea salt particles covering (a) nesting domain 3 and (b) Corsica. Violet line shows difference of reference scenarios. Blue line shows mean of difference of corresponding LD members. Blue contour covers the range of the standard deviation.

To determine the impact of sea salt particles, a visualization of the difference in the precipitation rate of the area-averaged value of REF ALL and

REF WO and of the mean of the corresponding LD members is given in Figure 6.10(a). Again, the standard deviation of LD is plotted as a blue contour. In both differences, there are positive and negative values; thus, that there is no signal showing that sea salt particles produce more or less rain on average at all times.

Differences in the precipitation rate are the same order of magnitude and are independent of the driving model. The standard deviation increases over time and has values of 0.05 to 0.075 mm/h. The spread generated by sea salt particles (visible in the difference) is as large as 25 to 37 % of the spread generated by the ensemble (visible in the absolute values).

The values of the standard deviation are larger than the absolute values of the difference, making it difficult to identify significant effects. Thus, taking the area average of such a large domain may cancel out important effects. I also took smaller domains (Figure 6.1) into account. As an example, the differences in the area-averaged precipitation rate over Corsica are plotted in Figure 6.10(b). Although the differences and standard deviation are larger over Corsica, the overall behavior is the same. The differences in the precipitation rate fluctuate by approximately 0 mm/h. The line is interrupted when there is no precipitation. Also, the standard deviation is larger than the absolute values, which is also true for the other sub-domains (not shown). Therefore, even if small areas are considered, taking the area average of a variable is an unfavorable approach for studying the impact of aerosols because significant effects may cancel out. Additionally, aerosols can have very localized impacts that are not visible in an area average. To overcome these disadvantages, representative data that take local features into account are needed. DAS meets these requirements.

6.2.1. Results of the Application of DAS

DAS is applied to the 3 h cumulative precipitation of each pair of corresponding LD members (with and without sea salt particles). Thus, 16 applications of DAS for each sub-domain (Figure 6.1), and threshold and maximum search distance combination are performed. The same values for the threshold and maximum search distance as in the 3 h cumulative precipitation of the reference scenario (subsection 6.1.1) are used: 5 mm, 10 mm, 20 mm, and 30 mm and a maximum search distance of 31.7 km.

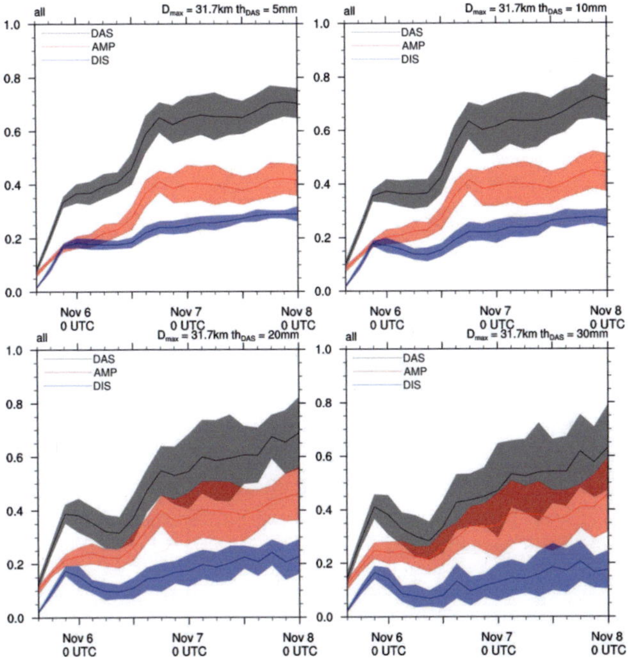

Figure 6.11.: Calculated mean of DAS (black line), AMP (red), and DIS (blue) for LD members and a constant maximum search distance of $D_{\text{max}} = 31.7\,\text{km}$ for 3 h cumulative precipitation over time. Shaded areas cover the range of mean value $\pm$ standard deviation. Used values for threshold th_{DAS} and maximum search distance D_{max} are found on the top right of each plot.

The mean values of DAS, AMP and DIS of the corresponding LEPS members over time are plotted as lines in Figure 6.11. The shaded areas cover the range of each mean value $\pm$ the corrected sample standard deviation (Equation 6.1). Compared with the scores of the reference scenarios (Figure 6.6), the overall behavior is quite similar. The contribution to the differences in the 3 h cumulative precipitation from different precipitation amounts is higher than the contribution from spatial shifts (the red lines are above the blue lines). Intersections only appear for $th_{\text{DAS}} = 5\,\text{mm}$ and $th_{\text{DAS}} = 10\,\text{mm}$ at approximately 18 UTC on November 5. The intersection for the reference scenario appears at the same time.

Over time, DAS increases, mainly because of the increasing AMP. In the reference scenarios (Figure 6.6), this behavior is step-wise (for thresholds of 5 mm and 10 mm). For small thresholds (5 mm and 10 mm), DIS also distinctly increases with the simulation time; for $th_{\mathrm{DAS}} = 20\,\mathrm{mm}$ and $th_{\mathrm{DAS}} = 30\,\mathrm{mm}$, the increase is smaller. The standard deviation of DIS is smaller than the standard deviation of AMP for all thresholds. As the threshold increases, the standard deviation of DIS increases. This finding is true for the standard deviation of AMP, but the growth is much smaller here.

Increasing the maximum search distance from 31.7 km to 63.4 km (not shown) results in findings that are similar to the reference scenarios. The overall behavior shows only slight differences between the applications of the two maximum search distances. Displacements in precipitation larger than 31.7 km ($F_{\mathrm{DAS}} = 1$) are thus not linked to sea salt particles. This result is also true for the sub-domains. Again, there are points in time when DAS is zero because of the lack of precipitation. The corresponding plots are shown in section C.2.

The findings achieved by the application of DAS to only one scenario pair (reference scenarios, see subsection 6.1.1) are not necessarily the impact of sea salt particles. Achieving the same results in several scenario pairs (LD, this section) demonstrates that the findings (DAS and AMP increase with time, increase in DIS is smaller, displacements are below 31.7 km) are robust and are statistically significant.

6.3. Analysis of the Initial Temperature Perturbation Ensemble

To separate the impact of sea salt particles from other factors, I created an ensemble using a small, random variation in the time derivative of the temperature at the first time step. Using the technique described in subsection 5.1.2, I created an ensemble with three members for the scenarios with and without sea salt particles. For the two reference scenarios without disturbance, the ensemble consists of eight members. Compared with the LD ensemble discussed in the previous section, pair building between corresponding members with and without sea salt particles is not possible. The random numbers used to generate the ensemble are seeded with the system clock of the super computer, which is different for each member. This ensemble has the name ITP (**I**nitial **T**emperature **P**erturbation) in the following.

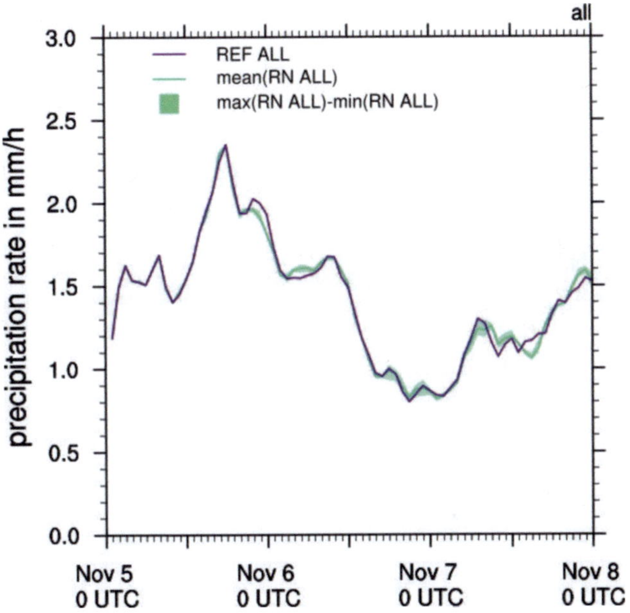

Figure 6.12.: Precipitation rates over time for scenarios taking sea salt particles into account covering nesting domain 3. Violet line shows scenario REF ALL. Turquoise line shows mean of ITP members with sea salt particles. Turquoise contour covers the spread calculated by the subtraction of the maximum and the minimum value of ITP ALL at each hour.

6.3.1. Results of the Application of FM

The impact of sea salt particles on precipitation is overlaid by the impact of other factors, as shown above (section 4.5). To quantify the impacts of each factor, the Factorial Method is applied as described in subsection 5.2.2. For the application, two factors with two values each are chosen. The result is a 2^2 FM design. The factors are the explicit treatment of the sea salt particles and the introduction of initial random perturbations of temperature (subsection 5.1.2). Both of the factors can be switched on or off (i.e., the two possible values). The calculation of the relative contribution of the factors to the total variance is achieved by dividing the sum of squares of each factor (Equation 5.22, Equation 5.23, Equation 5.24, and Equation 5.28) by the total sum of squares (Equation 5.26).

48 h Cumulative Precipitation

The relative contribution of sea salt particles, the random temperature perturbations, and their interaction to the variability in the cumulative precipitation over 48 h is calculated with FM starting on November 6 at 00 UTC. For the calculation, grid cells that have precipitation values below particular thresholds are neglected. The same thresholds used to calculate DAS for the reference scenarios (subsection 6.1.1) for the 48 h cumulative precipitation were assigned (25 mm, 50 mm, and 100 mm). Another very low threshold of 0.01 mm was used to exclude grid cells with no or very low precipitation. The contributions of the factors for each threshold are listed in Table 6.2.

Table 6.2.: Relative contribution of each factor and the interaction of the factors on differences in cumulative precipitation over 48 h.

relative	threshold			
contribution	0.01 mm	25 mm	50 mm	100 mm
sea salt	0.007	0.031	0.397	0.395
disturbance	0.028	0.007	0.200	0.012
interaction	0.795	0.000	0.013	0.214
error	0.170	0.962	0.390	0.379

For the minimum threshold of 0.01 mm, the interaction of sea salt particles and the disturbance is the major contributor to changes in the cumulative precipitation. Sea salt particles affect precipitation, but this effect cannot be separated from that of the random perturbation. A clear statement of the importance of sea salt particles for this threshold cannot be drawn. For thresholds of 50 mm and 100 mm, sea salt particles have the highest contribution, followed by the error contribution. The error can be interpreted as an uncertainty resulting from the disturbance because it originates from the different outcomes of the replicates. If each ensemble member of ITP produced the same final precipitation field, the error would be zero. The error is most prominent for a threshold of 25 mm. Here, all of the other factors have a very low contribution; in this special case, sea salt has no impact on the difference in the cumulative precipitation.

There is no link between the importance of the factors and the chosen thresholds, as seen in Table 6.2. The time span may be too long, or there may be no actual link. To account for the temporal changes, I present the application of FM to the 3 h cumulative precipitation for all time intervals in period 2 in the next section.

3 h Cumulative Precipitation

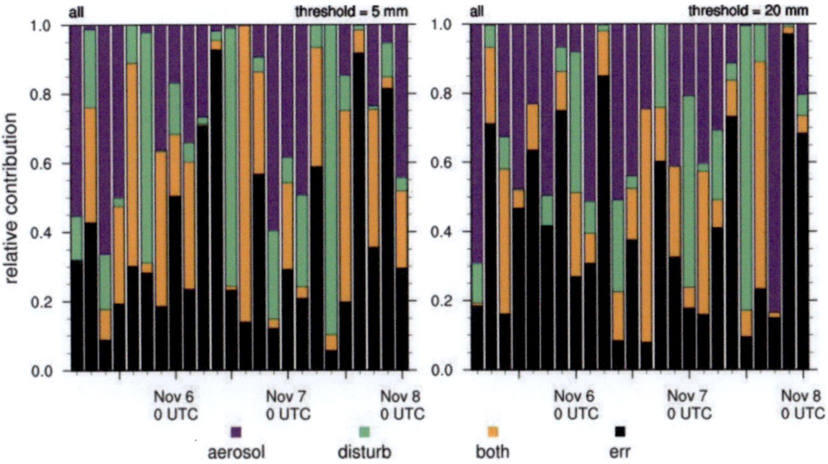

Figure 6.13.: Relative contribution of factors (sea salt particles in violet, disturbance in turquoise, the interaction of sea salt particles and disturbance in orange, and the error in black) to changes in 3 h cumulative precipitation over time. Grid cells showing a minimum cumulative precipitation of 5 mm (left plot) or 20 mm (right plot) over 3 h contribute to the calculation.

Figure 6.13 shows the relative contribution of each factor to the changes in the cumulative precipitation over 3 h for nested domain 3. The bars representing the relative contribution of each factor are stacked; to determine the contributions, the bottom value must be subtracted from the top value.

The left plot uses all grid cells that have at least 5 mm of precipitation over 3 h to calculate the relative contributions. This minimum value represents light rain, whereas the plot to the right represents moderate to heavy rain with a precipitation threshold of 20 mm. Because few grid cells show high precipitation, the number of grid cells that contribute to the calculation of FM in the right plot of Figure 6.13 is less than that in the left plot. Note that the contribution of the error is higher for a threshold of 20 mm on average.

Overall, none of the factors show a clear signal over time. For each individual time step, a different factor has the largest contribution to the changes in precipitation. A temporal signal, for example, an increasing contribution of the disturbance, is not visible.

Applying FM to the sub-domains shown in fig:sub-domains leads to the same conclusion. The plots of the sub-domain are within section C.3. Of course, fewer grid cells contribute to FM for the sub-domains compared with the entire nested domain 3. When the threshold of precipitation is set to 20 mm, even fewer grid cells remain, which means that the statement of FM cannot be easily generalized. Nevertheless, the contributions of the factors differ by time period without showing a trend, as is true for the nested domain 3 visible in Figure 6.13.

To discuss the impact of sea salt particles on precipitation, further constraints have to be implemented. The following section describes the application of FM to fields that show a displacement or differences in the amplitude of the precipitation field. Therefore, the fields achieved by the application of DAS (subsection 6.1.1) are used.

6.3.2. Results of the Application of the Coupling of DAS and FM

To separate the impact of sea salt particles on precipitation from those of other factors while concerning the type of change is not possible by the application of FM or DAS alone (see subsection 5.2.3). This section presents the combination of both methods and it's application on the precipitation field.

48 h Cumulative Precipitation

The relative contributions of sea salt particles, the disturbance and their interaction are calculated using FM with the 48 h cumulative precipitation starting on November 6 at 00 UTC. Grid cells, which show specific features achieved by DAS, contribute to FM. Thus, grid cells were filtered by those with a minimum value in $\mathrm{AMP}_{P1}(x, y)$, th_{AMP}, or in $\mathrm{DIS}_{P1}(x, y)$, th_{DIS}. The method was applied to several combinations of precipitation thresholds and maximum search distances, as achieved by the DAS calculations of the reference scenarios REF ALL and REF WO. The relative contributions over the minimum values are plotted in Figure 6.14.

The values for the threshold and maximum search distances used for the underlying DAS calculation are shown on the top right of each plot. The abscissa has staggered values. All of the grid cells that overcome the threshold of one of the factors (th_{AMP} and th_{DIS}) also contribute to the calculation of FM for lower thresholds of this factor. Thus, a the higher factor threshold corresponds to a lower number of contributing grid cells.

The four plots on top of Figure 6.14 show the relative contributions of the factors depending on the changes in the amplitude, as represented by th_{AMP}. The amplitude results from the root-mean-squared difference of precipitation after the morphing (for more information see subsection 5.2.1). Thus, the remaining precipitation difference after applying the shifts in the displacement field is given.

The top left image shows that sea salt particles contribute the most to changes in the precipitation for differences of at least 12 mm. For smaller differences, the contribution of the disturbance is greater. The orange line

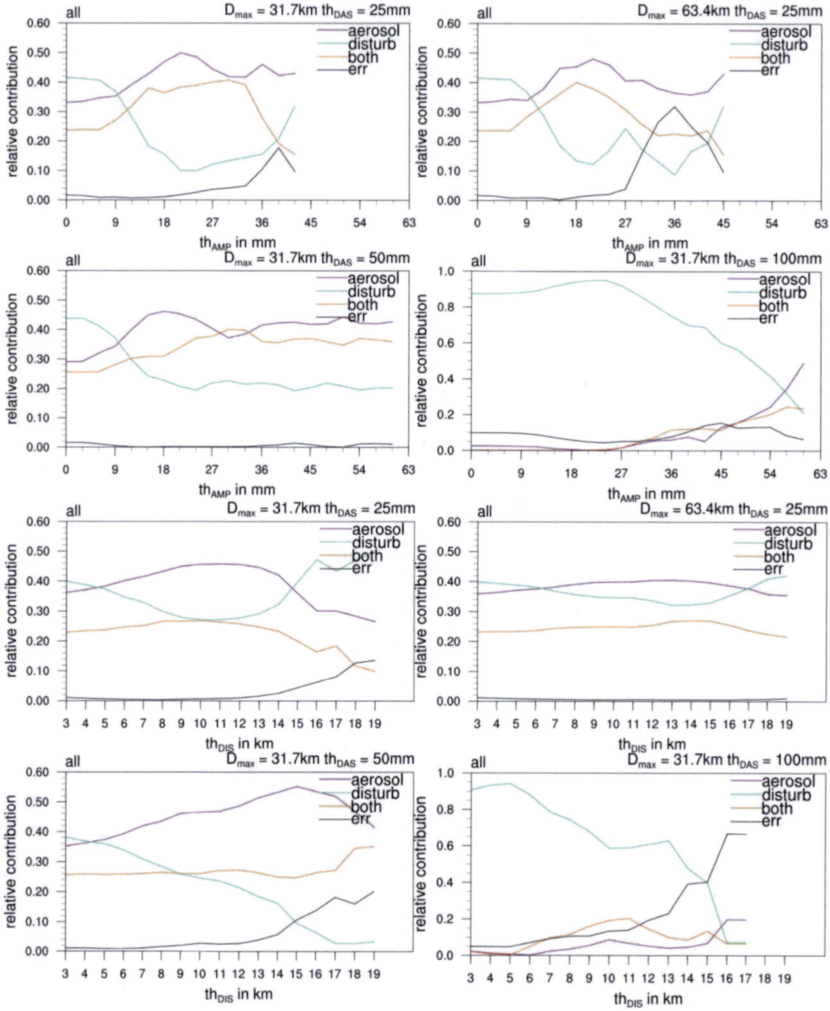

Figure 6.14.: Application of FM on DAS.

representing the contribution from the interaction of the sea salt particles and a disturbance is below of the aerosol line (violet) for each value in th_{AMP}. The error increases significantly for a minimum amplitude of

33 mm, because the number of grid cells strongly declines (not shown). Therefore, one should not interpret the values for amplitudes larger than 33 mm.

Increasing the maximum search distance for the same threshold also shows that sea salt particles are dominant at a minimum amplitude of 12 mm (top right). The relative contribution of up to 50 % is also the same as with the lower maximum search distance. For $D_{\mathrm{max}} = 63.4$ the error begins to increase at a minimum amplitude of 27 mm, which is lower than that for $D_{\mathrm{max}} = 31.7$ km. Specifically, the number of grid cells is lower for this th_{AMP} with the greater maximum search distance.

Changing the maximum search distance to 31.7 km but increasing the threshold to 50 mm (second row, left plot in Figure 6.14) reveals that for an amplitude of at least 12 mm, sea salt particles contribute the most to the changes in the precipitation. For minimum amplitudes of 30 mm and 33 mm, the contribution due to the interaction is higher than that due to sea salt particles, in contrast to $th_{\mathrm{DAS}} = 25$ mm. For all chosen values of th_{AMP}, the error is very low, although the number of grid cells contributing to the FM calculation is very low starting at 36 mm.

Choosing an even higher threshold of $th_{\mathrm{DAS}} = 100$ mm (second row, right plot) hardly impacts the sea salt particles. For all minimum amplitude values, the disturbance is the major contributor (often over 80 %). Here, the number of grid cells is low compared with the smaller thresholds. For 30 mm minimum amplitude, less than 500 grid cells contribute to the calculation of FM. Thus, one should not interpret the values for amplitudes larger than 30 mm. For very intense precipitation, the impact of sea salt particles is difficult to determine in the 48 h cumulative precipitation, possibly due to the high precipitation values around the Alps. Here, orographic uplift quickly induces high supersaturation. This condition may accelerate the growth of cloud drops to rain drops compared with non-orographic uplift. Thus, the amount of precipitation hardly differs, regardless of the number of CCN.

The four plots on the bottom of Figure 6.14 show the relative contributions of the factors to changes in the precipitation due to displacements, represented by th_{DIS}. The grid cells were filtered to overcome the value in th_{DIS} by contributing to the calculation of FM. Maximum search distances and thresholds for DAS are chosen that correspond to the four plots at the top of the figure.

Sea salt particles are most important for $D_{\max} = 31.7\,\text{km}$ and $th_{\text{DAS}} = 25\,\text{mm}$ for a minimum displacement of 6 km to 14 km. For smaller and larger minimum displacements, the disturbance is most important. The interaction of both factors is with approximately 25 % contribution less important than the factors alone. Starting at 13 km minimum displacement, the error rises remarkably.

Increasing the maximum search distance from 31.7 km to 63.4 km and keeping the threshold constant at 25 mm seems to flatten and stretch the curves (third row, plot at the right). The contribution of sea salt particles is greater than that of other factors between a minimum of 7 km and 17 km displacement. The relative contribution is approximately 40 % and is thus smaller than that for the maximum search distance of 31.7 km. Additionally, the contribution of the disturbance is greater and is therefore more similar to the contribution of sea salt particles. The interaction hardly differs compared to the smaller maximum search distance. The error is very low because the number of grid cells is above 4000 for displacements of 19 km. That displacements are less responsible for changes in precipitation can also be seen by the numbers of DIS in Table 6.1.

A decrease in the maximum search distance to 31.7 km but an increase in the threshold to 50 mm (last row on the left) also show that sea salt particles contribute the most to changes in precipitation due to shifts of at least 5 km. The influence of the disturbance decreases with increasing displacements without changing direction, in contrast to $th_{\text{DAS}} = 25\,\text{mm}$. The error starts to increase at a minimum displacement of 13 km, for which the number of contributing grid cells is approximately 500. Up to this minimum displacement, the relative contribution of the interaction is constant (approximately 26 %).

Increasing the threshold to 100 mm (bottom row, right plot) has nearly no impact of sea salt particles on changes of precipitation due to displacements, as is also true for changes due to the amplitude (second row, right). For a displacement of at least 4 km, less than 500 grid cells contribute to the calculation of FM. One should stop interpreting the values for larger displacements than 4 km. Here, the disturbance is dominant. High precipitation occurs around the Alps (see above) as a result of orographic uplift; thus, aerosols cannot cause displacements here because the precipitations is fixed to orography.

3 h Cumulative Precipitation

In Figure 6.15, the relative contributions of each factor to changes in precipitation by amplitude are plotted. Each factor is represented by a sub-figure denoted on the top left. All of the time intervals of the 3 h cumulative precipitation within period 2 are used, but information on the time scales is lost. There are 24 time intervals for the 3 h cumulative precipitation over a simulation period of 3 days. A specific DAS calculation with a maximum search distance of 31.7 km and a threshold of 5 mm underlie the FM calculation. The grid cells are filtered according to the thresholds in the amplitude field, th_{AMP}, plotted on the abscissa. The plot is a bivariate histogram showing the absolute number of time intervals for a specific combination of relative contributions and th_{AMP} in color. The black dots indicate the average relative contribution of the number of time intervals for each th_{AMP} value of the corresponding factor.

A larger th_{AMP} corresponds to fewer time intervals. Initially, a minimum amplitude of 4 mm has 24 time intervals, a 16 mm amplitude has 10 time intervals, and a 20 mm amplitude has only 5 time intervals. Thus, large differences in the amplitude occur at a few points in time.

On average, sea salt particles contribute the most to intensity changes in the precipitation for all chosen th_{AMP} values (top left). The contribution is mostly within the range of 35 to 40 %, but a few time intervals show contributions of 70 % and higher. The disturbance and its interaction with sea salt particles have nearly equal contributions. The error is constantly low, with average values up to 15 %; only single time intervals reach 55 %. For all th_{AMP}, the differences between the contributions of sea salt particles, the disturbance and interaction are very low. Thus, clear statements cannot be made in this presentation of data, although the contribution of sea salt particles is largest for all amplitudes.

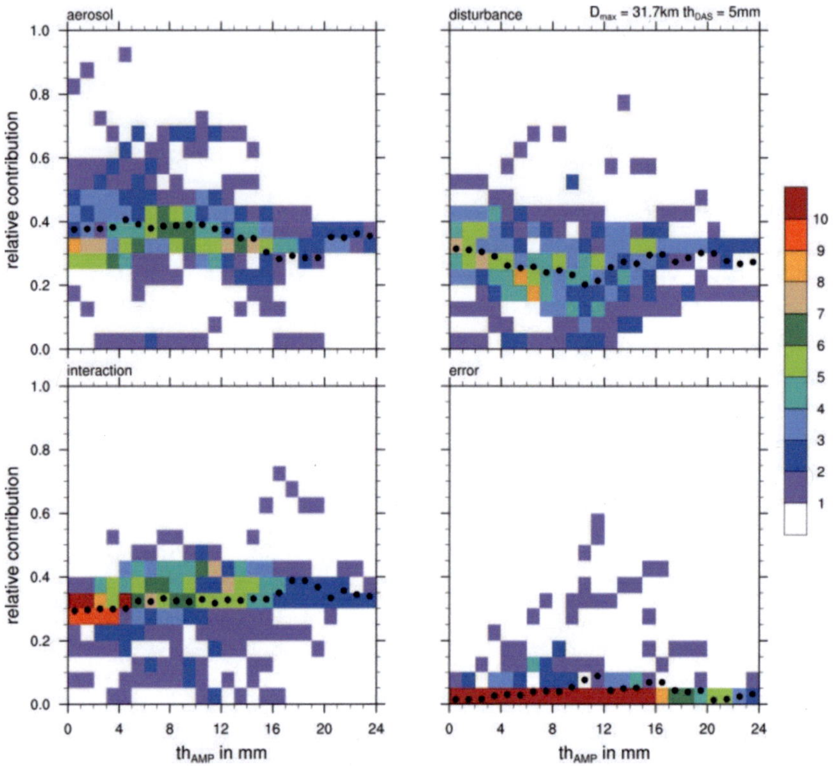

Figure 6.15.: Application of FM on several AMP fields for 3 h precipitation visualized as bivariate histograms. In color the absolute occurrence of time intervals is given. Black dots indicate the time intervals' average of relative contribution at specific th_{AMP}.

The contribution of the disturbance decreases from approximately 33 % for no minimum amplitude to 20 % for a minimum amplitude of 10 mm. Therefore, a disturbance leads to differences in the precipitation intensity of a few mm, regardless of the forecast lead time.

Overall, Figure 6.16 shows the same difference as Figure 6.15 except that the grid cells are filtered according to the th_{DIS} values rather than the th_{AMP} values. Even for large minimum displacements, most of the time intervals contribute to FM. For a minimum displacement of 20 km, there are still 16 time intervals available, but the number sharply decreases to one

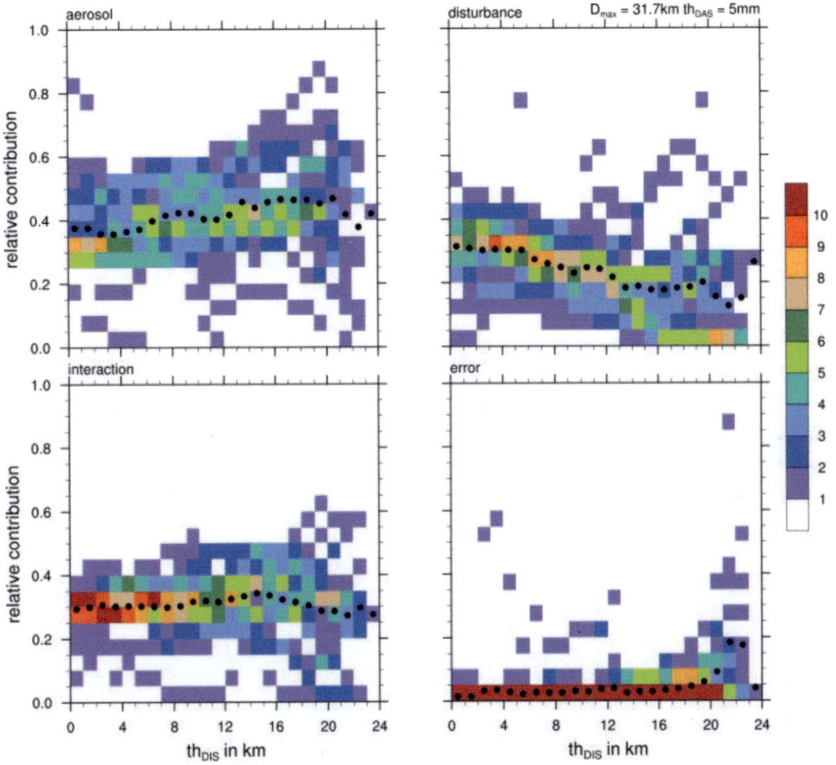

Figure 6.16.: Application of FM on several DIS fields for 3 h precipitation visualized as bivariate histograms. In color the absolute occurrence of time intervals is given. Black dots indicate the time intervals' average of relative contribution at specific th_{DIS}.

time interval for 23 km. Larger displacements do not occur, although the setting for the underlying DAS calculation with a maximum search distance of 31.7 km allows displacements of up to 31.7 km. Previously, displacements were below this value, but it is clearer here that displacements are mostly less than 20 km.

On average (black dots), sea salt particles contribute the most to shifts in 3 h precipitation (35 to 45 %). Compared with the differences in the precipitation amplitude, the discrepancy of the contributions of other factors is larger. The contribution of the error is low for all minimum displacements with an average below 10 %.

The contribution of the interaction is approximately 30 % with small fluctuations, and the contribution of the disturbance remarkably decreases as the minimum displacement increases. At the same time, the contributions of sea salt particles increase, particularly from a minimum displacement of 6 km (approximately 37 %) to 17 km (approximately 47 %). Later, the average contribution remains at a high level but with a decreasing number of time intervals.

An interesting feature in simulating aerosol effects is the time scale at which aerosol effects can be seen. As the simulation time increases, elucidating the impacts of aerosols becomes difficult because impacts due to nonlinear behavior grow rapidly. To determine these time scales, I used a combination of DAS and FM in several configurations, as shown in the following paragraphs.

The relative contributions to the 3 h cumulative precipitation over time are visualized in Figure 6.17. The contributing grid cells are filtered on the basis of the DAS calculations of the reference scenarios REF ALL and REF WO (see subsection 6.1.1); the maximum search distance and threshold are plotted on the top right of each sub-plot. Grid cells that show a minimum amplitude of the value plotted on the top left (th_{AMP}) based on the DAS calculations are taken into account for the resulting FM calculation.

Differences in the precipitation due to amplitudes induced by sea salt particles of at least 3 mm are visible as late as to 21 UTC on November 5, corresponding to a 21 h forecast (top left). Afterward, the contributions of sea salt particles, the disturbance and their interaction are balanced at approximately 30 %. In this case, sea salt particles are responsible for the precipitation differences in the form of intensity differences of at least 3 mm through the 21 h forecast. The error is only a few percent.

In the first few hours (up to 9 h), sea salt particles are not the main contributor. Additionally, the error is very high compared with the following hours. There are only a few grid cells that contribute to the calculation of FM because the atmospheric state is very similar in all scenarios. In this initialization phase, the model plateaus and significant impacts of specific factors are difficult to identify.

Increasing the minimum amplitude from 3 to 5 mm for the same DAS configuration (top right) shows a smaller time span in which sea salt particles contribute the most to changes in the precipitation due to intensity differences. Any significant impact would occur between the 9 and 12 h forecasts. Because of the small number of grid cells, the variation in the contributions by time step is high.

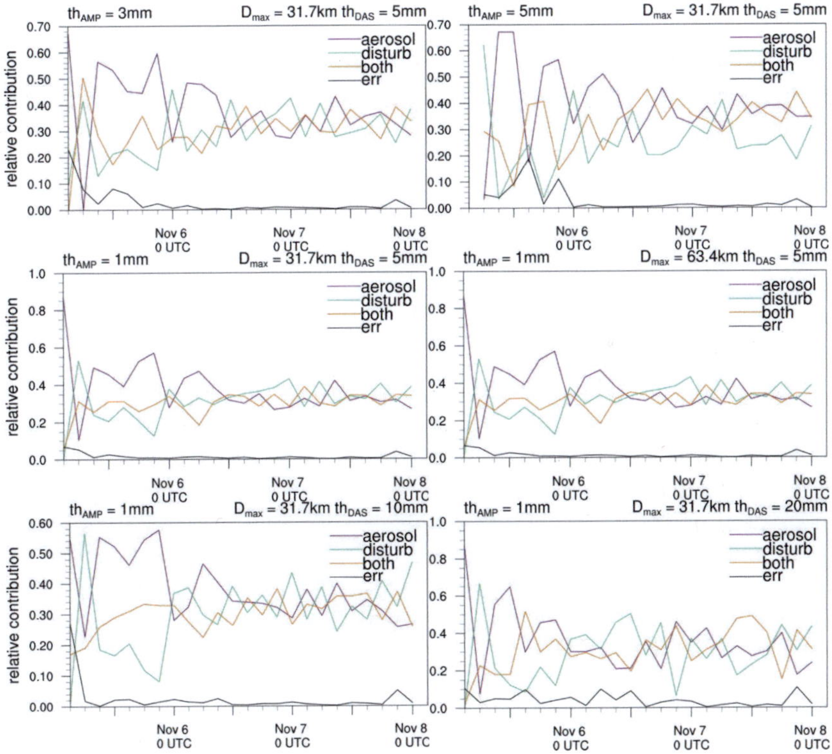

Figure 6.17.: Relative contribution of factors based on FM on differences in 3 h precipitation intensity for several amplitude thresholds over time. Underlying DAS configuration is plotted on the top right of each sub-plot.

Choosing a minimum amplitude of 1 mm (middle row, left) results in the same contribution of sea salt particles as the minimum of 3 mm; the contribution is the greatest between the 9 and 21 h forecasts. Afterward, all of the contributions are approximately the same size. If a larger maximum search distance is used for the underlying DAS calculation, then no difference is visible (middle row, right). This finding confirms that displacements occur below 31.7 km. Thus, the same corresponding precipitation objects are found regardless of the maximum search distances.

If the threshold for the DAS calculation increases from 5 to 10 mm (bottom left), then the time scale showing the impact of sea salt particles on precipitation is the same. When further increasing the threshold to 20 mm (bottom right), no time scale shows a significant impact of sea salt particles. First, the number of grid cells is too small to show significant effects; second, only locations with very high precipitation remain. As stated previously, this location is near the Alps, which is characterized by orographically induced precipitation; sea salt particles are not expected to have an impact in this area.

The relative contributions of each factor to precipitation changes due to spatial shifts over time are displayed in Figure 6.18. Several realizations of DAS are used as a basis. Note that results taken from the 48 h cumulative precipitation can differ from those taken from the 3 h cumulative precipitation. Thus, thresholds for the displacement that are independent of the findings of the 48 h cumulative precipitation are chosen. Choosing grid cells with a minimum displacement of 1 km (top left) shows that sea salt particles are dominant between 9 and 21 h. Larger minimum displacements of 8 km (middle left) or 12 km (top right) for the same DAS configuration show a contribution through 36 hours. Although fewer grid cells contribute here compared with a displacement of at least 1 km, these displacements are more important because the grid spacing is as large as 2.8 km; the corresponding resolution is even higher. Thus, it is possible to find displacements of 1 km, but they may not be linked to significant impacts.

Increasing the maximum search distance of DAS from 31.7 km to 63.4 km (middle right) hardly changes the contribution of sea salt particles. Sea salt particles are still the dominant factor for changes in precipitation due to displacements. Increasing the maximum search distance even further

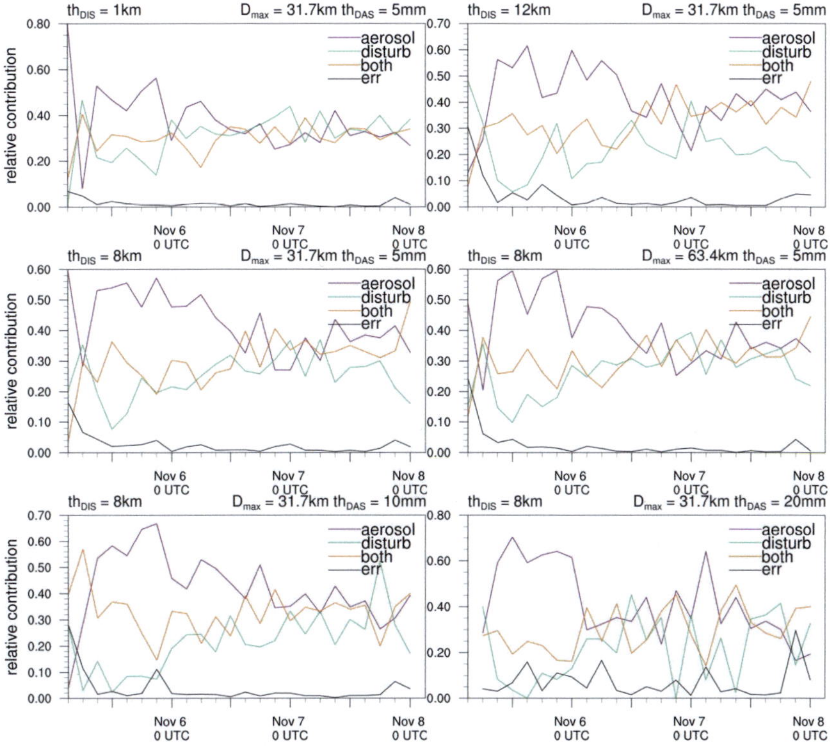

Figure 6.18.: Relative contribution of factors based on FM on shifts in 3 h precipitation for several displacement thresholds over time. Underlying DAS configuration is plotted on the top right of each sub-plot.

(not shown) does not change the contributions of the factors. Choosing minimum values of AMP (Figure 6.18) also confirms that displacements are smaller than 31.7 km.

Changing the threshold of DAS instead of the maximum search distance from 5 mm to 10 mm (bottom left) also indicates that sea salt particles are the main contributor in the 36 h forecast. Thus, for light and medium precipitation, an impact of sea salt particles on precipitation shifts can be seen for the first one and a half days of the simulation. For high precipitation ($th_{\mathrm{DAS}} = 20\,\mathrm{mm}$, bottom right), the time span is shorter (approximately one day). Here, because of the smaller number of grid cells, the error is large (up to 20 %). Compared with the differences in amplitude for $th_{\mathrm{DAS}} = 20\,\mathrm{mm}$, a significant impact is still visible.

In summary, the impact of sea salt particles on light and medium precipitation due to displacements is apparent for 1.5-day simulations. Differences in the amplitude are visible up to one simulated day. For intense precipitation, the impact of sea salt particles on displacements is apparent for a simulation time of one day, whereas the impact on the amplitude is not visible.

7. Conclusion

I identified a method for studying the feedbacks between aerosols, clouds and precipitation in the context of Medicanes. This method relies on an ensemble of model runs with different initial conditions to determine uncertainties in initial conditions. Small changes can rapidly lead to a different model response because of the non-linear atmospheric equations of motion (Lorenz, 1963). This effect must be separated from the effect of aerosols. The combination of the Displacement and Amplitude Score (Keil and Craig, 2007, 2009) and a Factorial Method (Teller and Levin, 2008) can extract the impact of aerosols on clouds and precipitation and can identify the type of impact.

Although much effort has been focused on gaining comprehensive insight into aerosol, cloud, and precipitation interactions in recent decades, there is still a large gap in the understanding of these processes (e.g., Tao et al., 2012). Model studies that account for individual cloud or weather systems are feasible for expanding this knowledge (e.g., Stevens and Feingold, 2009).

The interaction of aerosol, the winds, and precipitating clouds are studied under the dynamic conditions of a Medicane. This type of cyclone episodically occurs over the Mediterranean Sea (e.g. Cavicchia and von Storch, 2012; Tous and Romero, 2013) and resembles a small Hurricane (e.g., Billing et al., 1983; Ernst and Matson, 1983). Once a Medicane makes landfall, extensive damage is inevitable. Because of its high wind speeds, a Medicane induces large emission of sea salt particles, thus the cyclone creates its own aerosol field. If and how aerosols interact with a Medicane is the subject of this study.

I simulated a Medicane that occurred in early November 2011 using the modeling system COSMO-ART (Vogel et al., 2009). After confirming that the model system was able to represent the Medicane accurately, I created two aerosol scenarios concerning the representation of sea salt particles. Other types of aerosols, such as mineral dust or anthropogenic aerosols, were similarly represented in both scenarios. To account for the uncertainties in the initial conditions, inaccuracies due to (spatial and temporal) discretizations, closure assumptions in parameterizations, and effects of the numerical solver, these scenarios were depicted using two types of ensembles. In the first ensemble, I used the members of an already existing ensemble for the initial and boundary data to drive the simulations. In the second ensemble, I added small, random perturbations at the beginning of the simulation to the time derivative of the temperature. Without these ensembles, it would not be possible to determine the aerosol effect on clouds and precipitation due to interlaced effects, such as the non-linearity of dynamics.

The ensembles were the basis for the analysis. Many studies on the aerosol–cloud–precipitation feedback have been performed by comparing area-averaged precipitation values for different aerosol realizations. This approach, of course, provides no information on spatial distributions, nor does it account for buffering effects; thus, the effects will be misinterpreted according to Stevens and Feingold (2009). In this study, there are no significant differences in the area-averaged or area-accumulated precipitation for the different scenarios. Thus, a different method was applied for the first time. Classical grid-based scores are not capable of accounting for differences caused by displacements. A score that does not compare values of single corresponding grid cells is needed. To accomplish this task, the Displacement and Amplitude Score (Keil and Craig, 2007, 2009) was chosen, which compares two precipitation fields using an optical-flow technique and quantifies how well features are matched in terms of location and intensity. To extract the effect due to aerosols, the application of a Factorial Method (Teller and Levin, 2008) was performed. Finally, the combination of Displacement and Amplitude Score and the Factorial Method applied to the ensemble members can provide robust conclusions.

At the beginning of this work I formulated questions to address in this study. To conclude this work, I now return to these questions and provide explicit answers.

1. *How can aerosol effects on clouds be determined? Can a better method be developed for separating the effects of aerosol from those of the non-linear behavior of dynamics?*

   Single model runs are not sufficient to account for non-linearity; thus, ensembles with perturbations in the initial data are required. Based on the model signal, Factorial Methods can be used to determine single effects and the interaction of several effects. Grid-cell-based scores are poor when the focus is highly localized quantities, such as cloud cover or precipitation. The Displacement and Amplitude Score is a feasible measure for studying localized features.

2. *What are the effects of sea salt particles on a Medicane, particularly its precipitation?*

   Neither the sum nor the area average of precipitation displayed noteworthy differences in the different aerosol realizations. In contrast, local high differences were visible. Areas with high or low precipitation values could be distinguished at various time scales. Sea salt particles seemed to have no significant impact on the overall structure or the track of the Medicane.

   - *Do particles lead to spatial and temporal shifts in precipitation because the onset of precipitation formation is delayed or accelerated? Do sea salt particles cause changes in the precipitation amount?*

     The particles have an impact on both the location and intensity of precipitation. Depending on the time scale and region, one of the factors is dominant. For example, shifts occurred in the 48 h cumulative precipitation along the French coast. However, in all of the 3 h cumulative precipitation fragments, intensity differences commonly occurred everywhere. Typical differences in the intensity reached 12 mm within 3 h.

   - *How large are spatial shifts in precipitation? How fast does the impact of sea salt particles on precipitation propagate (i.e. how large is this signal speed)? Is the signal speed coupled to the horizontal wind velocity, the Medicane's transition velocity, or neither?*

     Regardless of the time scales, spatial shifts are mainly less than 31.7 km. These shifts can be attributed to the impact of sea salt particles when the shifts are at least 6 km. The upper bound depends on the precipitation's intensity and the time scales. In the 48 h cumulative precipitation case, shifts caused by sea salt particles last up to 14 km for 25 mm to 50 mm cumulative precipitation. For stronger precipitation, sea salt particles are responsible for shifts up to 19 km. In the 3 h cumulative precipitation, a shift of up to 20 km is caused by sea salt particles for less intense precipitation. At both time scales, sea salt particles do not cause shifts in very high precipitation. The signal speed of the impact of sea salt particles on precipitation was approximately 3 m/s. This speed does not depend on the transition velocity or the horizontal wind speeds, since both of which both would result in much higher signal speeds.

- *On what time scales can aerosol effects be observed in the simulations? Is aerosol observability dependent on the lifetime of single clouds (which would be minutes to a few hours) or on the lifetime of the entire system (a few days)?*

  The time scales at which effects of aerosols on precipitation can be seen depend on the intensity of precipitation. Impacts on the location of precipitation objects are found up to a 36 h lead time, whereas impacts on the amplitude of the precipitation are found up to a 24 h lead time. Thus, the analysis using the proposed method indicates that the time scales do not depend on the lifetime of single clouds or on the lifetime of the Medicane itself.

The combination of the Displacement and Amplitude Score and a Factorial Method can determine the impact of aerosols on clouds and precipitation and the type of impact. In principle, this method is applicable to other variables and weather phenomena and thus is a proper tool for studying aerosol feedbacks to clouds and precipitation.

A. Initial Aerosol

In COSMO-ART the size distributions of aerosol are represented by modes. The number and mass of each mode is described by log-normal distributions. As soot is part of the mixed modes, they have at least a diameter of the diameter of soot. The median diameters of number distributions differs from those of mass distributions. The standard deviation is the same for both, the number and the mass distribution. The values used to display the size distribution for number concentrations of the modes in Figure 3.2 are listed in Table A.1.

Table A.1.: Aerosol data for Figure 3.2

Mode i	standard devi. σ_i	median diam. $\overline{d_{p,N,i}}$ in µm	number conc. N_i in cm^{-3}
nucleation	1.7	0.01	100,000
nucleation mixed	1.7	0.08	10,000
accumulation	2.0	0.07	5,000
accumulation mixed	2.0	0.08	5,000
soot	1.4	0.08	2,000
coarse	2.5	1.00	10
dust A	1.7	0.64	500
dust B	1.6	3.50	8
dust C	1.5	8.70	2
sea salt A	1.9	0.20	500
sea salt B	2.0	2.00	5
sea salt C	1.7	12.00	2

Values for the dimensionless standard deviation are kept constant within COSMO-ART. Values for median diameters are also the initial values in COSMO-ART which are used in the calculation of emissions. In REF WO there is no calculation of sea salt emission, thus there are no sea salt particles within this scenario. Values for the number concentrations are chosen to build a hypothetically environment over Mediterranean coastal regions.

B. Gedankenexperiment of the Factorial Method

For better understanding of FM, here comes a small theoretical example of a 2^2 factorial design with three replicates. Therefore, 12 experiment runs are necessary. This example is taken from Montgomery (2008). Table B.1 gives the outcome of each of the replicates and it's total (Equation 5.17) for all treatment combinations.

Table B.1.: Data of the gedankenexperiment

Treatment Combination	Replicate I	II	III	Total
$\Psi(a_1, b_1)$	28	25	27	80
$\Psi(a_2, b_1)$	36	32	32	100
$\Psi(a_1, b_2)$	18	19	23	60
$\Psi(a_2, b_2)$	31	30	29	90

The effect of A, B, and the interaction of both can be calculated by Equation 5.18, Equation 5.20 and Equation 5.21, respectively

$$\begin{aligned}
\mathcal{A} &= \frac{1}{2N}\left(\Psi(a_2,b_1)+\Psi(a_2,b_2)-\Psi(a_1,b_2)-\Psi(a_1,b_1)\right) \\
&= \frac{1}{2\cdot 3}(100+90-60-80) && = 8.33, \\
\mathcal{B} &= \frac{1}{2N}\left(\Psi(a_2,b_2)+\Psi(a_1,b_2)-\Psi(a_2,b_1)-\Psi(a_1,b_1)\right) \\
&= \frac{1}{2\cdot 3}(90+60-100-80) && = -5.00, \\
\mathcal{AB} &= \frac{1}{2N}\left(\Psi(a_2,b_2)+\Psi(a_1,b_1)-\Psi(a_2,b_1)-\Psi(a_1,b_2)\right) \\
&= \frac{1}{2\cdot 3}(90+80-100-60) && = 1.67.
\end{aligned}$$

The absolute values of the effects suggest, that the effect of the interaction $\mathcal{AB}$ is small in comparison to the effects $\mathcal{A}$ and $\mathcal{B}$. Increasing factor A results in an increased outcome of the experiment, whereas an increase of factor B decreases the experiment's result.

To confirm these first guesses, one can calculate the sum of squares for the factors A and B and the interaction of both by Equation 5.22, Equation 5.23 and Equation 5.24

$$\begin{aligned}
SS_A &= \frac{\left(\Psi(a_2,b_2)+\Psi(a_2,b_1)-\Psi(a_1,b_2)-\Psi(a_1,b_1)\right)^2}{4N} \\
&= \frac{(50)^2}{4\cdot 3} && = 208.33, \\
SS_B &= \frac{\left(\Psi(a_2,b_2)+\Psi(a_1,b_2)-\Psi(a_2,b_1)-\Psi(a_1,b_1)\right)^2}{4N} \\
&= \frac{(-30)^2}{4\cdot 3} && = 75.00, \\
SS_{AB} &= \frac{\left(\Psi(a_2,b_2)+\Psi(a_1,b_1)-\Psi(a_2,b_1)-\Psi(a_1,b_2)\right)^2}{4N} \\
&= \frac{(10)^2}{4\cdot 3} && = 8.33.
\end{aligned}$$

The total sum of squares SS_T is calculated by Equation 5.26

$$\begin{aligned} SS_T = & \sum_{i=1}^{n} \left(\Psi_i(a_2,b_1)^2 + \Psi_i(a_1,b_2)^2 + \Psi_i(a_2,b_2)^2 + \Psi_i(a_1,b_1)^2\right) \\ & - \frac{\left(\Psi(a_2,b_1) + \Psi(a_2,b_2) + \Psi(a_1,b_2) + \Psi(a_1,b_1)\right)^2}{4N} \\ = & \; 9398.00 - 9075.00 = 323.00. \end{aligned}$$

The missing term is the sum of squares due to error. This is achieved by rearranging of Equation 5.27

$$\begin{aligned} SS_E &= SS_T - SS_A - SS_B - SS_{AB} \\ &= 323.00 - 208.33 - 75.00 - 8.33 \\ &= 31.34. \end{aligned}$$

This confirms the first guess, that the main effects are statistically significant (they are at least one order of magnitude larger then the interaction effect) and that there is no interaction between factors A and B. Also the error is of no importance.

C. Analysis of Sub-Domains

C.1. DAS for Reference Scenarios on 3 h Precipitation

The following plots are an extension to subsection 6.1.1. They show the values of DAS, AMP, and DIS for 3 h cumulative precipitation over time. Calculations are done separately for each sub-domain (see Figure 6.1). If DAS is zero it either means a complete agreement of the two reference scenarios in precipitation or, more likely, that there were less precipitation than the chosen threshold value (plotted on the top right of each plot).

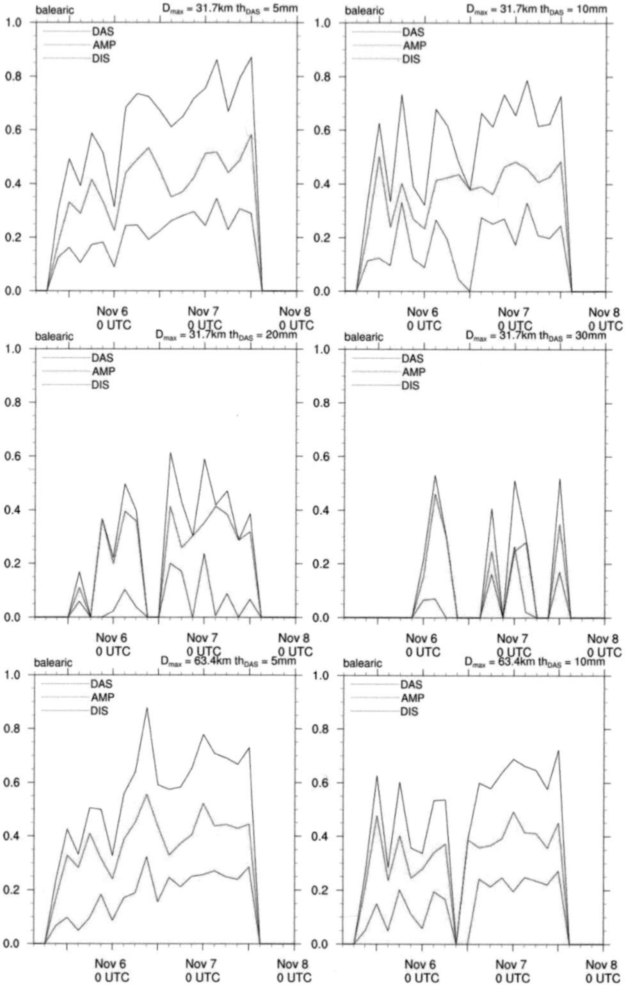

Figure C.1.: DAS (black line), AMP (red), and DIS (blue) for sub-domain 1 for 3 h cumulative precipitation over time. Values used for threshold th_{DAS} and maximum search distance $D_{\max}$ are noted on the top right of each plot.

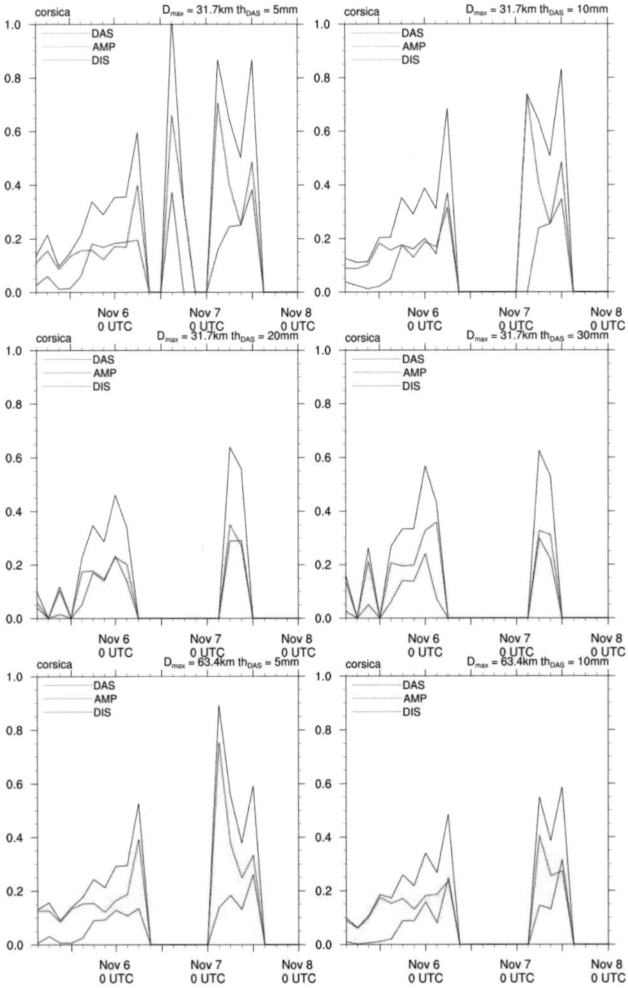

Figure C.2.: Same as Figure C.1 but for sub-domain 2.

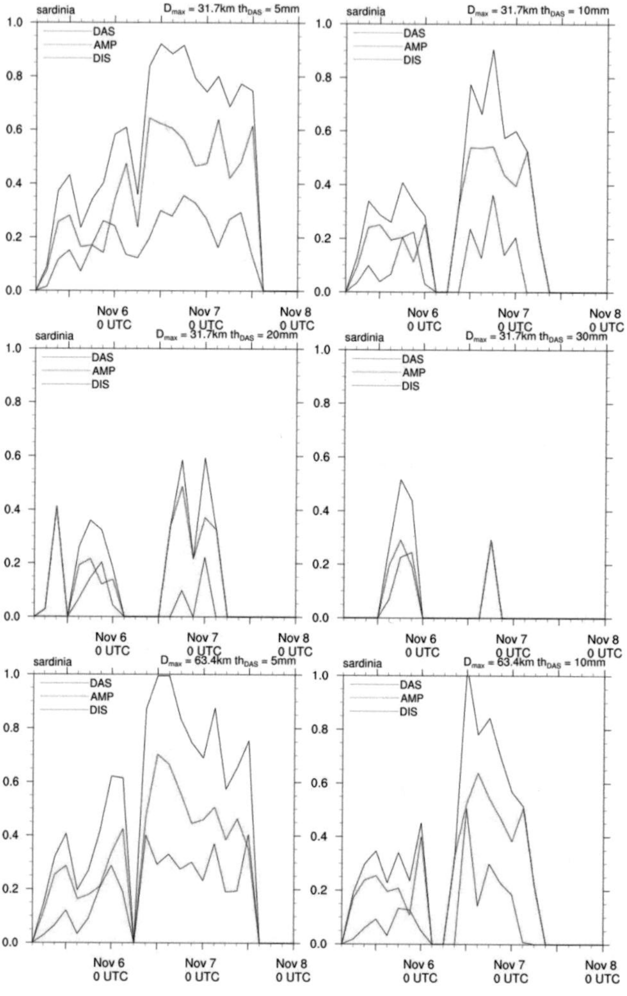

Figure C.3.: Same as Figure C.1 but for sub-domain 3.

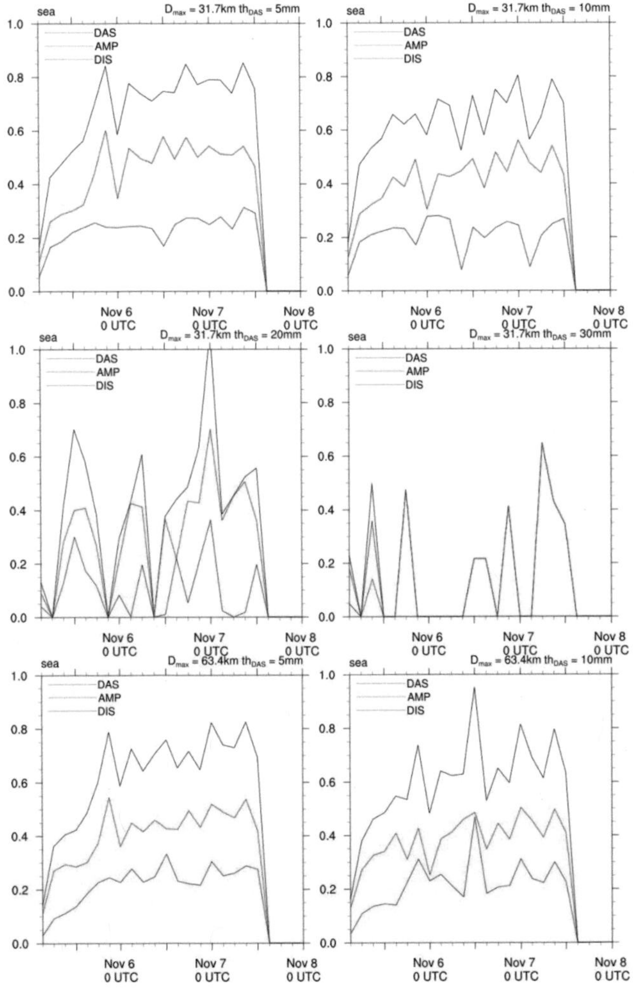

Figure C.4.: Same as Figure C.1 but for sub-domain 4.

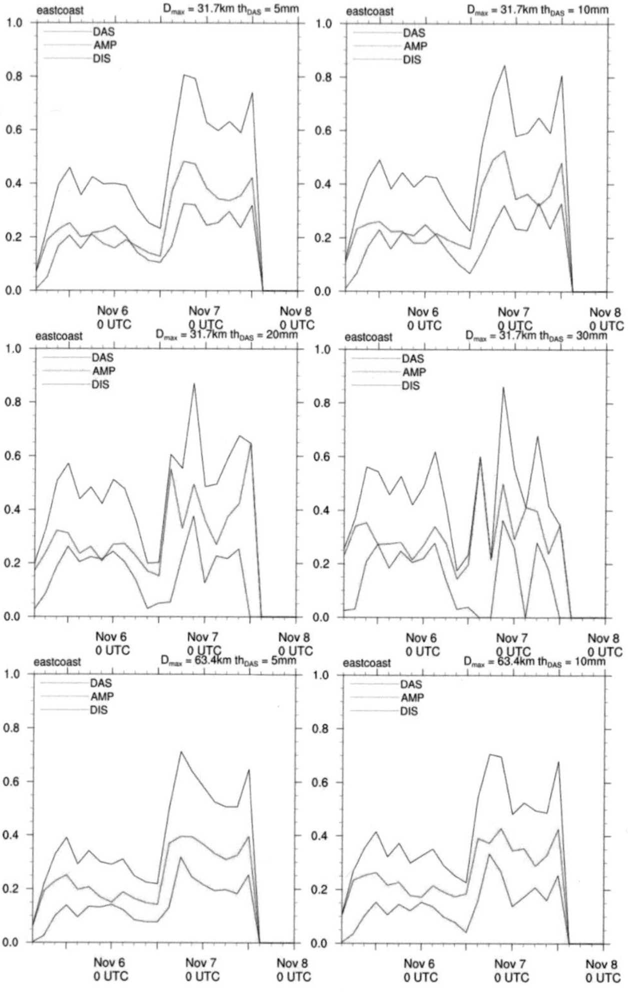

Figure C.5.: Same as Figure C.1 but for sub-domain 5.

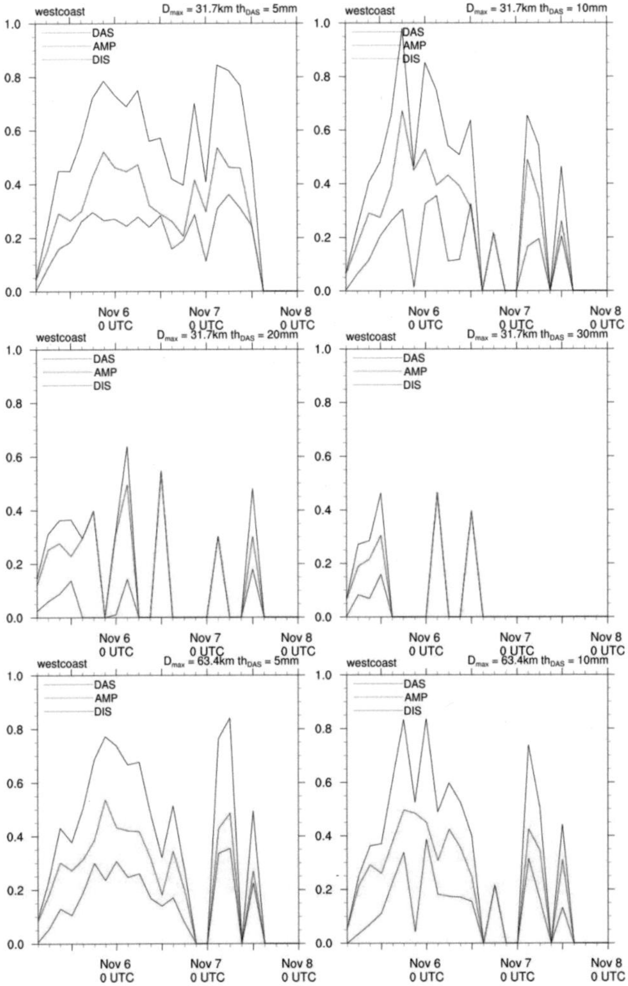

Figure C.6.: Same as Figure C.1 but for sub-domain 6.

C.2. DAS for LEPS Driven Ensemble on 3 h Precipitation

The following plots are an extension to subsection 6.2.1. They show the mean values of the LD members for DAS, AMP, and DIS for 3 h cumulative precipitation over time as solid lines. Calculations are done separately for each sub-domain (see Figure 6.1). If DAS is zero it either means a complete agreement of each of the two corresponding scenarios for all members in precipitation or, more likely, that there were less precipitation than the chosen threshold value (plotted on the top right of each plot) in all members. Shaded areas cover the range of mean values ± the standard deviation calculated with Equation 6.1. If there is no shading visible, just one of the members contribute to the standard deviation's calculation which results in a zero value.

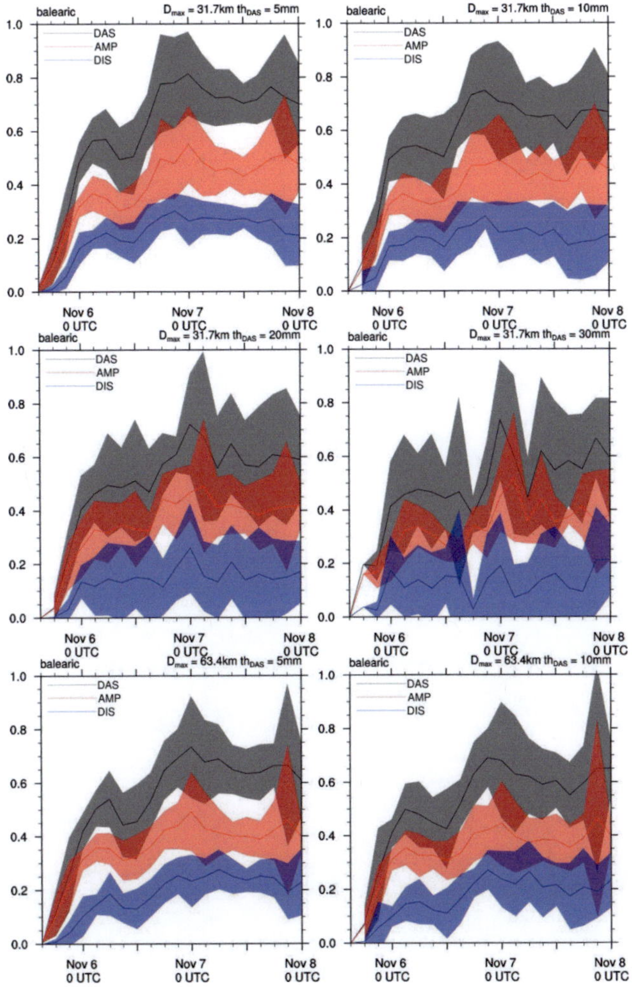

Figure C.7.: LD ensemble mean of DAS (black line), AMP (red), and DIS (blue) for sub-domain 1 for 3 h cumulative precipitation over time. Shaded areas cover the range of mean value ± standard deviation. Values used for threshold th_{DAS} and maximum search distance $D_{\max}$ are noted on the top right of each plot.

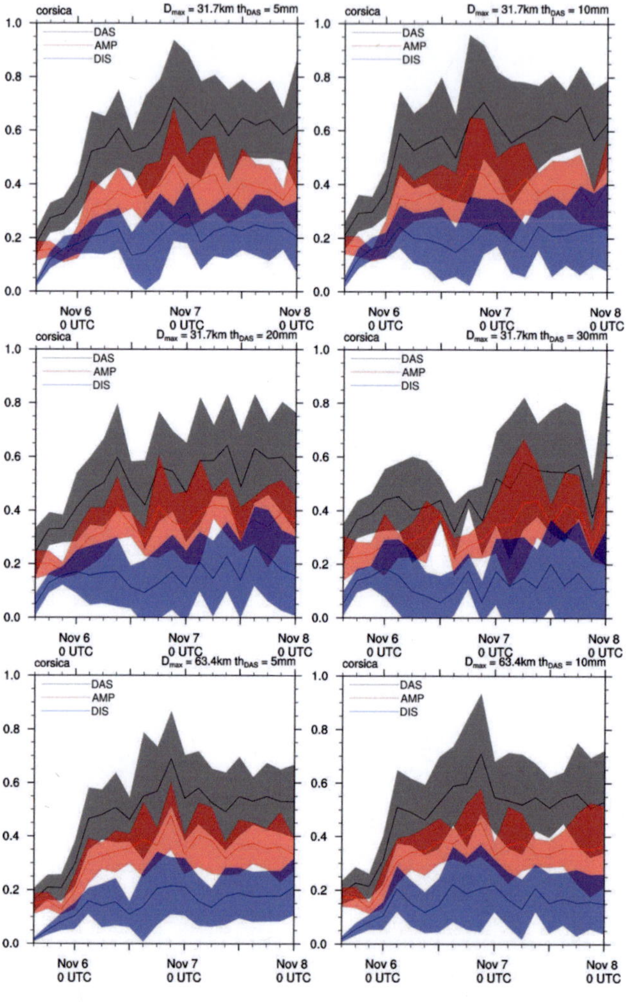

Figure C.8.: Same as Figure C.7 but for sub-domain 2.

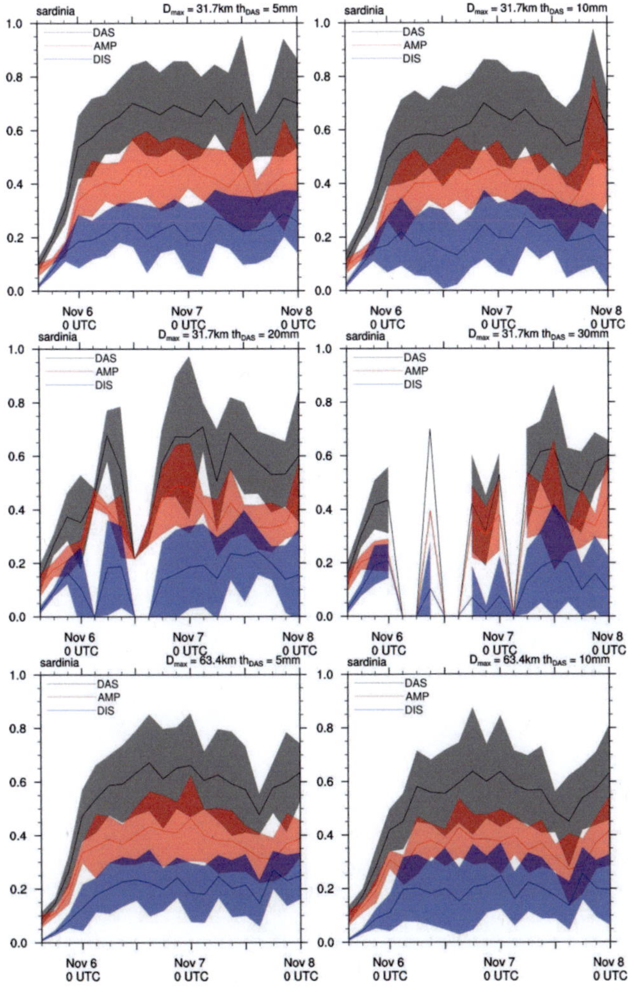

Figure C.9.: Same as Figure C.7 but for sub-domain 3.

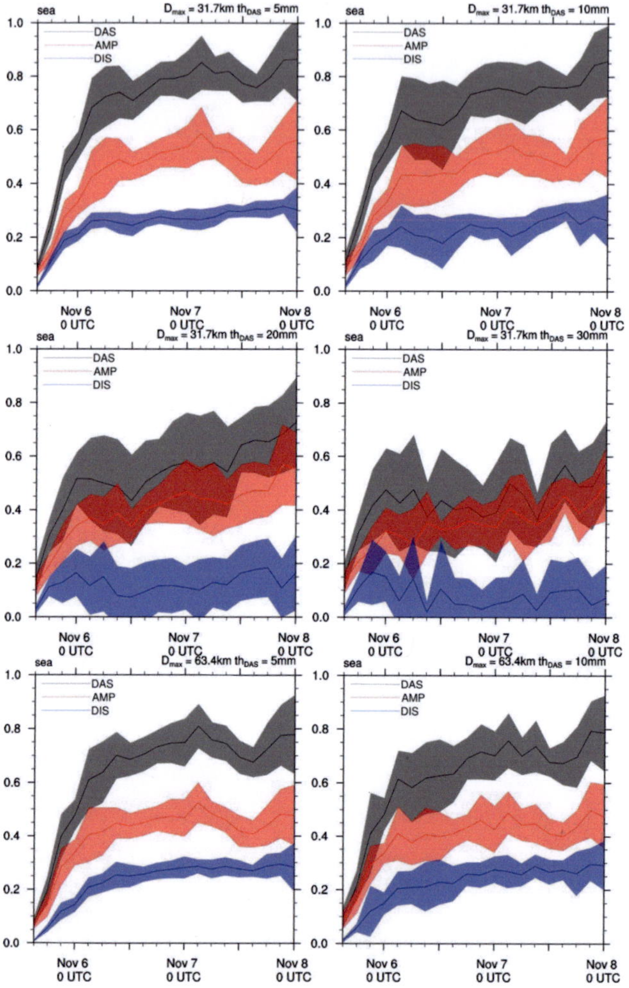

Figure C.10.: Same as Figure C.7 but for sub-domain 4.

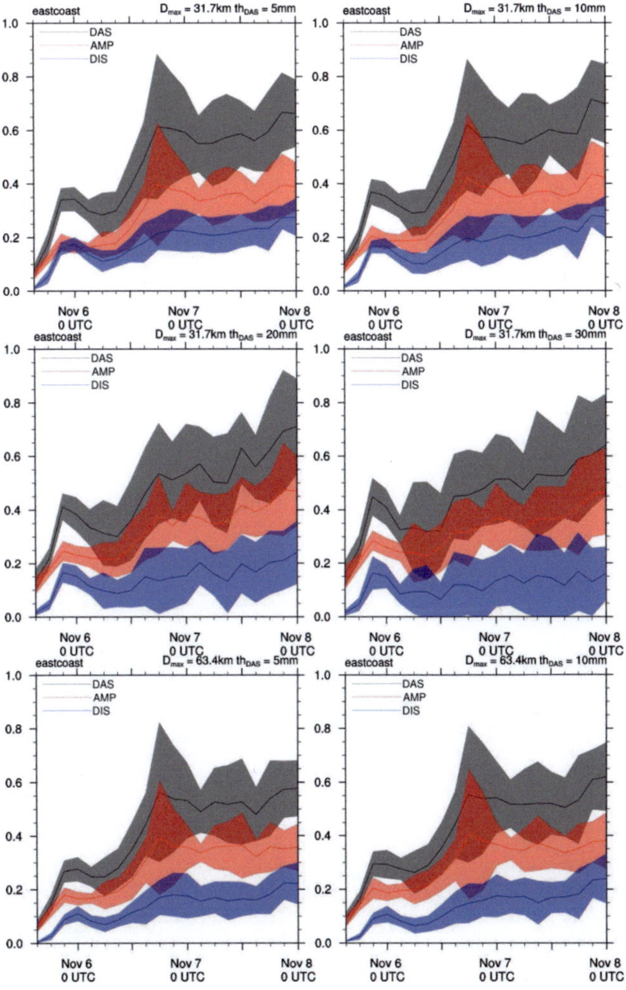

Figure C.11.: Same as Figure C.7 but for sub-domain 5.

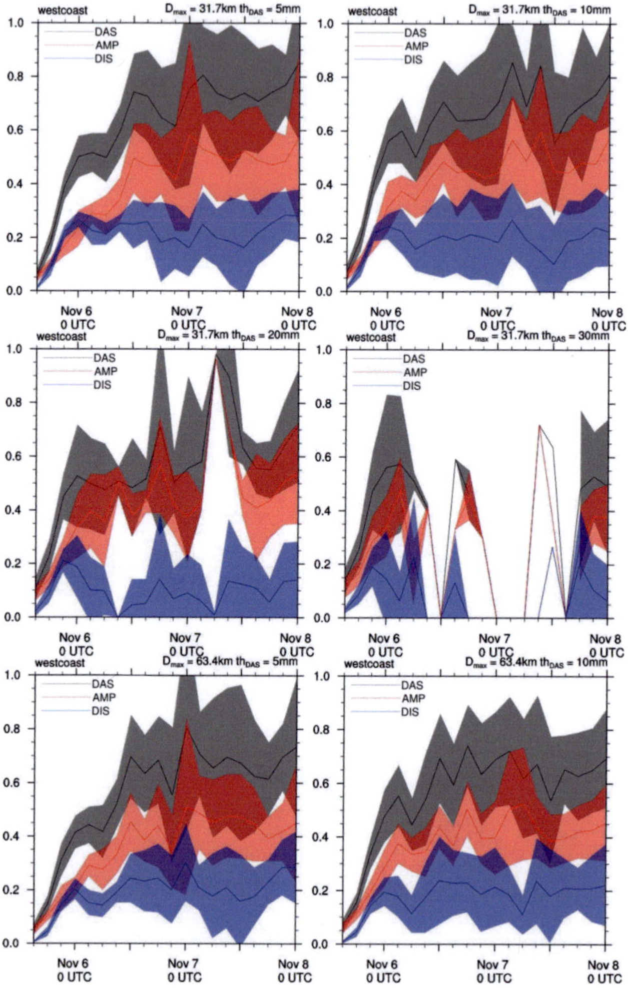

Figure C.12.: Same as Figure C.7 but for sub-domain 6.

C.3. FM for Initial Temperature Perturbation Ensemble on 3 h Precipitation

The following plots are an extension to subsection 6.3.1. FM is applied to the 3 h cumulative precipitation over time on the sub-domains noted on the top left of each plot. Points in time with no bars plotted means that there are too less grid cells exceeding the threshold in precipitation (top right of each plot) to apply the FM calculation.

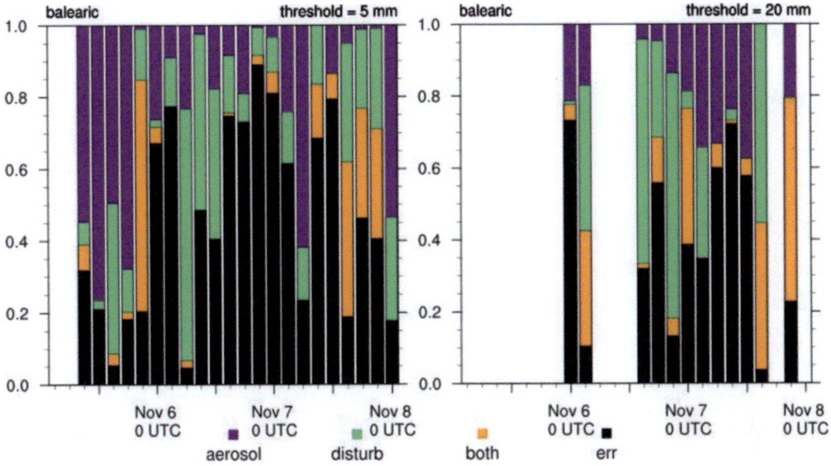

Figure C.13.: Relative contribution of factors to changes in 3 h cumulative precipitation over time for sub-domain 1. Grid cells showing a minimum cumulative precipitation of 5 mm (left plot) or 20 mm (right plot) over 3 h contribute to the calculation.

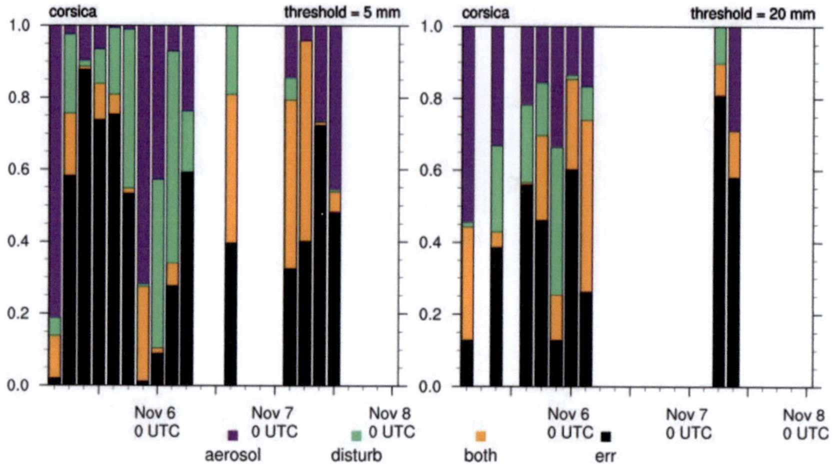

Figure C.14.: Same as Figure C.13 but for sub-domain 2.

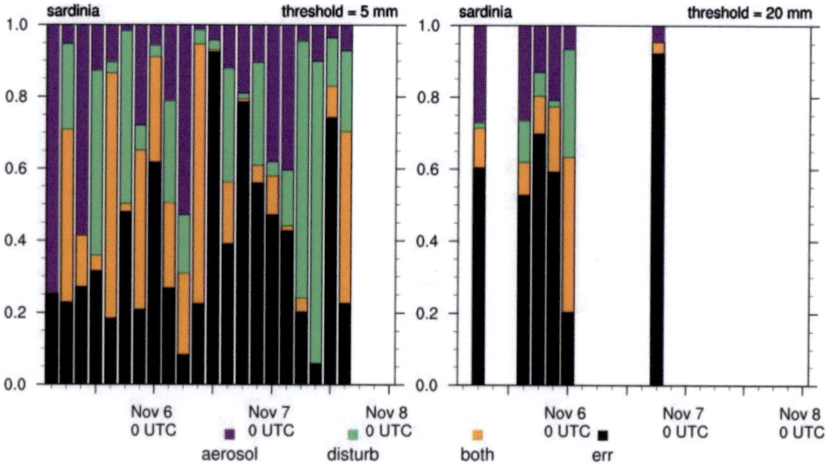

Figure C.15.: Same as Figure C.13 but for sub-domain 3.

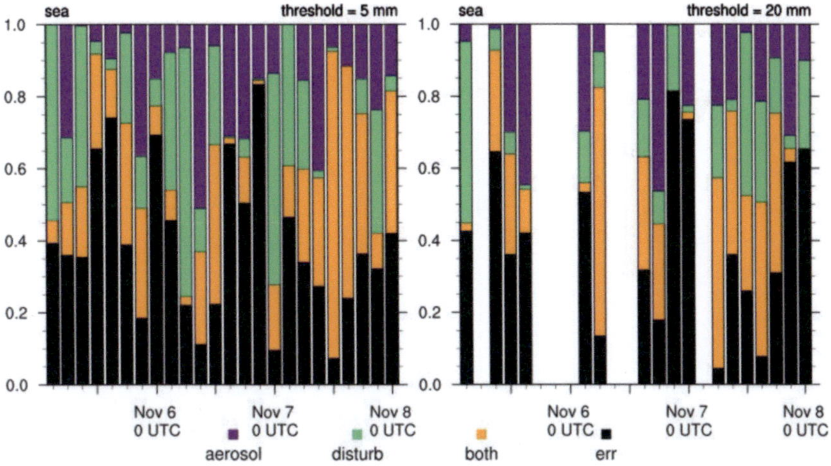

Figure C.16.: Same as Figure C.13 but for sub-domain 4.

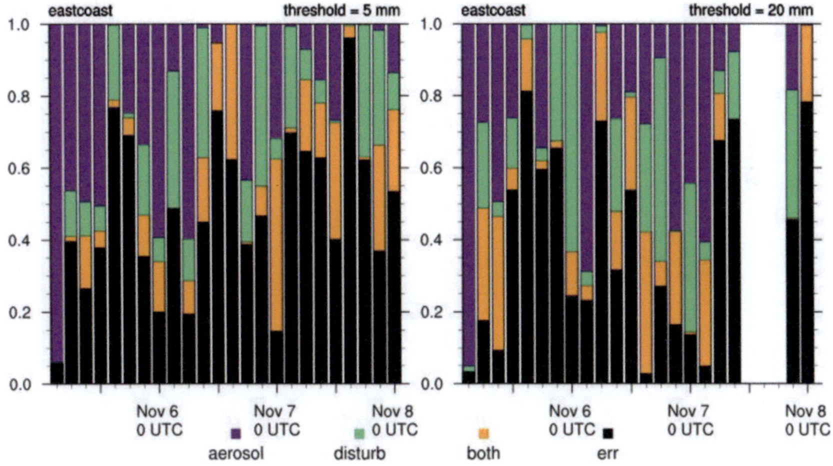

Figure C.17.: Same as Figure C.13 but for sub-domain 5.

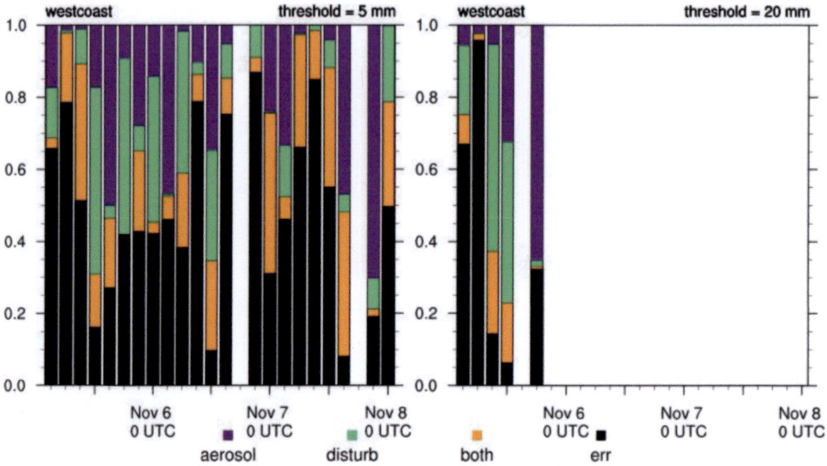

Figure C.18.: Same as Figure C.13 but for sub-domain 6.

List of Figures

List of Tables

Bibliography

Ackerman, A. S., O. Toon, D. Stevens, A. Heymsfield, V. Ramanathan, and E. Welton, 2000: Reduction of tropical cloudiness by soot. *Science*, **288 (5468)**, 1042–1047.

Ahijevych, D., E. Gilleland, B. G. Brown, and E. E. Ebert, 2009: Application of spatial verification methods to idealized and NWP-gridded precipitation forecasts. *Weather Forecast.*, **24 (6)**, 1485–1497.

Akhtar, N., J. Brauch, A. Dobler, K. Béranger, and B. Ahrens, 2014: Medicanes in an ocean–atmosphere coupled regional climate model. *Nat. Hazard. Earth Sys. Discuss.*, **2 (3)**, 2117–2149.

Albrecht, B. A., 1989: Aerosols, cloud microphysics, and fractional cloudiness. *Science*, **245 (4923)**, 1227–1230.

Alfaro, S. C., and L. Gomes, 2001: Modeling mineral aerosol production by wind erosion: Emission intensities and aerosol size distributions in source areas. *J. Geophys. Res.-Atmos.*, **106 (D16)**, 18 075–18 084.

Alpert, P., and T. Sholokhman, 2011: *Factor Separation in the Atmosphere: Applications and Future Prospects.* Cambridge University Press, Cambridge, United Kingdom, 292 p.

Andreae, M. O., D. Rosenfeld, P. Artaxo, A. Costa, G. Frank, K. Longo, and M. Silva-Dias, 2004: Smoking rain clouds over the Amazon. *Science*, **303 (5662)**, 1337–1342.

Athanasopoulou, E., H. Vogel, B. Vogel, A. Tsimpidi, S. Pandis, C. Knote, and C. Fountoukis, 2013: Modeling the meteorological and chemical effects of secondary organic aerosols during an EUCAARI campaign. *Atmos. Chem. Phys.*, **13 (2)**, 625–645.

Baklanov, A., and Coauthors, 2014: Online coupled regional meteorology chemistry models in Europe: current status and prospects. *Atmos. Chem. Phys.*, **14 (1)**, 317–398.

Baldauf, M., A. Seifert, J. Förstner, D. Majewski, M. Raschendorfer, and T. Reinhardt, 2011: Operational convective-scale numerical weather prediction with the COSMO model: description and sensitivities. *Mon. Weather Rev.*, **139 (12)**, 3887–3905.

Bangert, M., C. Kottmeier, B. Vogel, and H. Vogel, 2011: Regional scale effects of the aerosol cloud interaction simulated with an online coupled comprehensive chemistry model. *Atmos. Chem. Phys.*, **11 (9)**, 4411–4423.

Bangert, M., and Coauthors, 2012: Saharan dust event impacts on cloud formation and radiation over Western Europe. *Atmos. Chem. Phys.*, **12 (9)**, 4045–4063.

Barahona, D., and A. Nenes, 2009: Parameterizing the competition between homogeneous and heterogeneous freezing in ice cloud formation–polydisperse ice nuclei. *Atmos. Chem. Phys.*, **9 (16)**, 5933–5948.

Barahona, D., J. Rodriguez, and A. Nenes, 2010a: Sensitivity of the global distribution of cirrus ice crystal concentration to heterogeneous freezing. *J. Geophys. Res.-Atmos.*, **115 (D23)**.

Barahona, D., R. West, P. Stier, S. Romakkaniemi, H. Kokkola, and A. Nenes, 2010b: Comprehensively accounting for the effect of giant CCN in cloud activation parameterizations. *Atmos. Chem. Phys.*, **10 (5)**, 2467–2473.

Barnaba, F., and G. P. Gobbi, 2004: Aerosol seasonal variability over the Mediterranean region and relative impact of maritime, continental and Saharan dust particles over the basin from MODIS data in the year 2001. *Atmos. Chem. Phys.*, **4 (9/10)**, 2367–2391.

Behrendt, R., and H. Mahlke, 2011: wetter3.de Website. http://www1.wetter3.de. Accessed: 2014-09-16.

Billing, H., I. Haupt, and W. Tonn, 1983: Evolution of a hurricane-like cyclone in the Mediterranean Sea. *Beitr. Phys. Atm.*, **56 (4)**, 508–510.

Bott, A., 1989: A positive definite advection scheme obtained by nonlinear renormalization of the advective fluxes. *Mon. Weather Rev.*, **117 (5)**, 1006–1016.

Buizza, R., M. Leutbecher, and L. Isaksen, 2008: Potential use of an ensemble of analyses in the ECMWF Ensemble Prediction System. *Q. J. Roy. Meteor. Soc.*, **134 (637)**, 2051–2066.

Cavicchia, L., and H. von Storch, 2012: The simulation of medicanes in a high-resolution regional climate model. *Clim. Dynam.*, **39 (9-10)**, 2273–2290.

Cavicchia, L., H. von Storch, and S. Gualdi, 2014: A long-term climatology of medicanes. *Clim. Dynam.*, 1–13.

Chaboureau, J.-P., O. Nuissier, and C. Claud, 2012a: Verification of ensemble forecasts of Mediterranean high-impact weather events against satellite observations. *Nat. Hazard. Earth. Sys.*, **12 (8)**, 2449–2462.

Chaboureau, J.-P., F. Pantillon, D. Lambert, E. Richard, and C. Claud, 2012b: Tropical transition of a Mediterranean storm by jet crossing. *Q. J. Roy. Meteor. Soc.*, **138 (664)**, 596–611.

Conte, D., M. M. Miglietta, and V. Levizzani, 2011: Analysis of instability indices during the development of a Mediterranean tropical-like cyclone using MSG-SEVIRI products and the LAPS model. *Atmos. Res.*, **101 (1)**, 264–279.

Damian, V., A. Sandu, M. Damian, F. Potra, and G. R. Carmichael, 2002: The kinetic preprocessor KPP-a software environment for solving chemical kinetics. *Comput. Chem. Eng.*, **26 (11)**, 1567–1579.

Doms, G., and Coauthors, 2011: A Description of the Nonhydrostatic Regional COSMO-Model Part II: Physical Parameterization. Tech. rep., Deutscher Wetterdienst.

Drobinski, P., and Coauthors, 2014: HyMeX: a 10-year multidisciplinary program on the mediterranean water cycle. *B. A.M. Meteorol. Soc.*, **95 (7)**, 1063–1082.

Dulac, F., and Coauthors, 2009: The Chemistry-Aerosol Mediterranean Experiment (ChArMEx). *1st International Workshop, Toulouse, France, 29 June –1 July.*

Dunion, J. P., and C. S. Velden, 2004: The impact of the Saharan air layer on Atlantic tropical cyclone activity. *B. A.M. Meteorol. Soc.*, **85 (3)**, 353–365.

Dunstone, N., D. Smith, B. Booth, L. Hermanson, and R. Eade, 2013: Anthropogenic aerosol forcing of Atlantic tropical storms. *Nat. Geosci.*, **6 (7)**, 534–539.

Ebert, E. E., U. Damrath, W. Wergen, and M. E. Baldwin, 2003: The WGNE assessment of short-term quantitative precipitation forecasts. *B. A.M. Meteorol. Soc.*, **84 (4)**, 481–492.

Ebuchi, N., H. C. Graber, and M. J. Caruso, 2002: Evaluation of wind vectors observed by QuikSCAT/SeaWinds using ocean buoy data. *J. Atmos. Ocean. Tech.*, **19 (12)**, 2049–2062.

Ekman, A., A. Engström, and C. Wang, 2007: The effect of aerosol composition and concentration on the development and anvil properties of a continental deep convective cloud. *Q. J. Roy. Meteor. Soc.*, **133 (627)**, 1439–1452.

Ekman, A. M., C. Wang, J. Ström, and R. Krejci, 2006: Explicit simulation of aerosol physics in a cloud-resolving model: Aerosol transport and processing in the free troposphere. *J. Atmos. Sci.*, **63 (2)**, 682–696.

Emanuel, K., 2003: Tropical cyclones. *Annu. Rev. Earth. Pl. Sc.*, **31 (1)**, 75–104.

Emanuel, K., 2005: Genesis and maintenance of "Mediterranean hurricanes". *Adv. Geosci.*, **2 (2)**, 217–220.

Ernst, J., and M. Matson, 1983: A Mediterranean tropical storm? *Weather*, **38 (11)**, 332–337.

Evan, A. T., J. Dunion, J. A. Foley, A. K. Heidinger, and C. S. Velden, 2006: New evidence for a relationship between Atlantic tropical cyclone activity and African dust outbreaks. *Geophys. Res. Lett.*, **33 (19)**.

Eyre, J., 1991: A fast radiative transfer model for satellite sounding systems. *ECMWF Technical Memorandum*, **176**.

Fan, J., J. M. Comstock, M. Ovchinnikov, S. A. McFarlane, G. McFarquhar, and G. Allen, 2010: Tropical anvil characteristics and water vapor of the tropical tropopause layer: Impact of heterogeneous and homogeneous freezing parameterizations. *J. Geophys. Res.-Atmos.*, **115 (D12)**.

Feser, F., and M. Barcikowska, 2012: The influence of spectral nudging on typhoon formation in regional climate models. *Environ. Res. Lett.*, **7 (1)**, 014024.

Fita, L., R. Romero, A. Luque, K. Emanuel, and C. Ramis, 2007: Analysis of the environments of seven Mediterranean tropical-like storms using an axisymmetric, nonhydrostatic, cloud resolving model. *Nat. Hazard. Earth. Sys.*, **7 (1)**, 41–56.

Fita, L., R. Romero, A. Luque, and C. Ramis, 2009: Effects of assimilating precipitation zones derived from satellite and lightning data on numerical simulations of tropical-like Mediterranean storms. *Annales Geophysicae*, **27**, 3297–3319.

Fountoukis, C., and A. Nenes, 2005: Continued development of a cloud droplet formation parameterization for global climate models. *J. Geophys. Res.-Atmos.*, **110 (D11)**.

Fountoukis, C., and A. Nenes, 2007: ISORROPIA II: a computationally efficient thermodynamic equilibrium model for K+–Ca 2+–Mg 2+–NH 4+–Na+–SO 4 2—NO 3—Cl—H 2 O aerosols. *Atmos. Chem. Phys.*, **7 (17)**, 4639–4659.

Geiger, H., I. Barnes, I. Bejan, T. Benter, and M. Spittler, 2003: The tropospheric degradation of isoprene: an updated module for the regional atmospheric chemistry mechanism. *Atmos. Environ.*, **37 (11)**, 1503–1519.

Grell, G. A., and S. R. Freitas, 2014: A scale and aerosol aware stochastic convective parameterization for weather and air quality modeling. *Atmos. Chem. Phys.*, **14 (10)**, 5233–5250.

Grimmett, G., and D. Stirzaker, 2001: *Probability and Random Processes*. Probability and Random Processes, OUP Oxford, Oxford, Great Britain, 596 p.

Grup de Meteorologia UIB, 2007: Medicanes: Mediterranean tropical-like storms. http://www.uib.cat/depart/dfs/meteorologia/METEOROLOGIA/MEDICANES/. Accessed: 2014-09-16.

Hart, R. E., 2003: A cyclone phase space derived from thermal wind and thermal asymmetry. *Mon. Weather Rev.*, **131 (4)**, 585–616.

Hazra, A., P. Mukhopadhyay, S. Taraphdar, J.-P. Chen, and W. R. Cotton, 2013: Impact of aerosols on tropical cyclones: An investigation using convection-permitting model simulation. *J. Geophys. Res.-Atmos.*, **118 (13)**, 7157–7168.

Heymsfield, A. J., L. M. Miloshevich, C. Schmitt, A. Bansemer, C. Twohy, M. R. Poellot, A. Fridlind, and H. Gerber, 2005: Homogeneous ice nucleation in subtropical and tropical convection and its influence on cirrus anvil microphysics. *J. Atmos. Sci.*, **62 (1)**, 41–64.

Homar, V., R. Romero, D. Stensrud, C. Ramis, and S. Alonso, 2003: Numerical diagnosis of a small, quasi-tropical cyclone over the western Mediterranean: Dynamical vs. boundary factors. *Q. J. Roy. Meteor. Soc.*, **129 (590)**, 1469–1490.

Hoose, C., J. Kristjánsson, and S. Burrows, 2010: How important is biological ice nucleation in clouds on a global scale? *Environ. Res. Lett.*, **5 (2)**, 024 009.

Hu, Y., and K. Stamnes, 1993: An accurate parameterization of the radiative properties of water clouds suitable for use in climate models. *J. Climate*, **6 (4)**, 728–742.

Hummel, M., and Coauthors, 2014: Regional-scale simulations of fungal spore aerosols using an emission parameterization adapted to local measurements of fluorescent biological aerosol particles. *Atmos. Chem. Phys. Discuss.*, **14 (7)**, 9903–9950.

Igel, A. L., M. R. Igel, and S. C. van den Heever, 2015: Make it a double? Sobering results from simulations using single-moment microphysics schemes. *J. Atmos. Sci.*, **72 (2)**, 910–925.

Isono, K., M. Komabayasi, and A. Ono, 1959: The nature and origin of ice nuclei in the atmosphere. *J. Meteor. Soc. Japan*, **37 (6)**, 211–233.

Jiang, H., H. Xue, A. Teller, G. Feingold, and Z. Levin, 2006: Aerosol effects on the lifetime of shallow cumulus. *Geophys. Res. Lett.*, **33 (14)**.

Jiang, J. H., H. Su, M. R. Schoeberl, S. T. Massie, P. Colarco, S. Platnick, and N. J. Livesey, 2008: Clean and polluted clouds: Relationships among pollution, ice clouds, and precipitation in South America. *Geophys. Res. Lett.*, **35 (14)**.

Kain, J. S., and J. M. Fritsche, 1993: Convective parameterization for mesoscale models: The Kain–Fritsch scheme. *The representation of cumulus convection in numerical models, Meteor. Monogr*, **46**, 165–170.

Kallos, G., M. Astitha, P. Katsafados, and C. Spyrou, 2007: Long-range transport of anthropogenically and naturally produced particulate matter in the Mediterranean and North Atlantic: current state of knowledge. *J. Appl. Meteorol. Clim.*, **46 (8)**, 1230–1251.

Keil, C., and G. C. Craig, 2007: A displacement-based error measure applied in a regional ensemble forecasting system. *Mon. Weather Rev.*, **135 (9)**, 3248–3259.

Keil, C., and G. C. Craig, 2009: A displacement and amplitude score employing an optical flow technique. *Weather Forecast.*, **24 (5)**, 1297–1308.

Keil, C., A. Tafferner, and T. Reinhardt, 2006: Synthetic satellite imagery in the Lokal-Modell. *Atmos. Res.*, **82 (1)**, 19–25.

Khain, A., N. BenMoshe, and A. Pokrovsky, 2008: Factors determining the impact of aerosols on surface precipitation from clouds: An attempt at classification. *J. Atmos. Sci.*, **65 (6)**, 1721–1748.

Khain, A., A. Pokrovsky, M. Pinsky, A. Seifert, and V. Phillips, 2004: Simulation of effects of atmospheric aerosols on deep turbulent convective clouds using a spectral microphysics mixed-phase cumulus cloud model. Part I: Model description and possible applications. *J. Atmos. Sci.*, **61 (24)**, 2963–2982.

Khain, A., D. Rosenfeld, and A. Pokrovsky, 2005: Aerosol impact on the dynamics and microphysics of deep convective clouds. *Q. J. Roy. Meteor. Soc.*, **131 (611)**, 2639–2663.

King, M. D., Y. J. Kaufman, W. Menzel, and D. Tanre, 1992: Remote sensing of cloud, aerosol, and water vapor properties from the Moderate Resolution Imaging Spectrometer (MODIS). *Ieee. T. Geosci. Remote.*, **30 (1)**, 2–27.

King, N., K. Bower, J. Crosier, and I. Crawford, 2013: Evaluating MODIS cloud retrievals with in situ observations from VOCALS-REx. *Atmos. Chem. Phys.*, **13 (1)**, 191–209.

Knippertz, P., and J.-B. W. Stuut, 2014: *Mineral Dust: A Key Player in the Earth System.* Springer, Heidelberg New York London, 509 p.

Knote, C., and Coauthors, 2011: Towards an online-coupled chemistry-climate model: evaluation of trace gases and aerosols in COSMO-ART. *Geosci. Model Dev.*, **4 (4)**, 1077–1102.

Kottmeier, C., P. Palacio-Sese, N. Kalthoff, U. Corsmeier, and F. Fiedler, 2000: Sea breezes and coastal jets in southeastern Spain. *Geosci. Model Dev.*, **20 (14)**, 1791–1808.

Kraut, I., 2011: Numerische Simulation zum Einfluss von Halogenverbindungen auf die troposhärische Photooxidantienbildung auf der regionalen Skala, Diploma Thesis (KIT).

Kugler, U., J. Theloke, B. Thiruchittampalam, R. Friedrich, S. Laufer, M. Knecht, C. Schieberle, and T. Kampffmeyer, 2013: Projekt Verkehrsentwicklung und Umwelt (VEU), Abschlussdokument.

Kumar, P., I. Sokolik, and A. Nenes, 2009: Parameterization of cloud droplet formation for global and regional models: including adsorption activation from insoluble CCN. *Atmos. Chem. Phys.*, **9 (7)**, 2517–2532.

Lee, S. S., L. J. Donner, and V. T. Phillips, 2009: Impacts of aerosol chemical composition on microphysics and precipitation in deep convection. *Atmos. Res.*, **94 (2)**, 220–237.

Levin, Z., and W. R. Cotton, 2008: *Aerosol pollution impact on precipitation: a scientific review.* Springer, 407 p.

Loeb, N. G., and G. L. Schuster, 2008: An observational study of the relationship between cloud, aerosol and meteorology in broken low-level cloud conditions. *J. Geophys. Res.-Atmos.*, **113 (D14)**.

Lohmann, U., and J. Feichter, 2005: Global indirect aerosol effects: a review. *Atmos. Chem. Phys.*, **5 (3)**, 715–737.

Lorenz, E. N., 1963: Deterministic nonperiodic flow. *J. Atmos. Sci.*, **20 (2)**, 130–141.

Lundgren, K., B. Vogel, H. Vogel, and C. Kottmeier, 2013: Direct radiative effects of sea salt for the Mediterranean region under conditions of low to moderate wind speeds. *J. Geophys. Res.-Atmos.*, **118 (4)**, 1906–1923.

Luque, A., L. Fita, R. Romero, and S. Alonso, 2007: Tropical-like Mediterranean storms: an analysis from satellite. *EUMETSAT 07 proceedings.*

Lyamani, H., A. Valenzuela, D. Perez-Ramirez, C. Toledano, M. J. Granados-Muñoz, F. J. Olmo, and L. Alados-Arboledas, 2015: Aerosol properties over the western Mediterranean basin: temporal and spatial variability. *Atmos. Chem. Phys.*, **15 (5)**, 2473–2486.

Maheras, P., E. Xoplaki, T. Davies, J. Martin-Vide, M. Bariendos, and M. J. Alcoforado, 1999: Warm and cold monthly anomalies across the Mediterranean basin and their relationship with circulation; 1860–1990. *Geosci. Model Dev.*, **19 (15)**, 1697–1715.

Majewski, D., and Coauthors, 2002: The operational global icosahedral-hexagonal gridpoint model GME: Description and high-resolution tests. *Mon. Weather Rev.*, **130 (2)**, 319–338.

Marsigli, C., F. Boccanera, A. Montani, T. Paccagnella, and Coauthors, 2005: The COSMO-LEPS mesoscale ensemble system: validation of the methodology and verification. *Nonlinear Proc. Geoph.*, **12 (4)**, 527–536.

Mårtensson, E., E. Nilsson, G. de Leeuw, L. Cohen, and H.-C. Hansson, 2003: Laboratory simulations and parameterization of the primary marine aerosol production. *J. Geophys. Res.-Atmos.*, **108 (D9)**.

Mather, G., D. Terblanche, F. Steffens, and L. Fletcher, 1997: Results of the South African cloud-seeding experiments using hygroscopic flares. *J. Appl. Meteorol.*, **36 (11)**, 1433–1447.

Mathieu Barbery, 2011: Meteo60.fr Website. http://www.meteo60.fr. Accessed: 2014-09-16.

Miglietta, M., S. Laviola, A. Malvaldi, D. Conte, V. Levizzani, and C. Price, 2013: Analysis of tropical-like cyclones over the Mediterranean Sea through a combined modeling and satellite approach. *Geophys. Res. Lett.*, **40 (10)**, 2400–2405.

Miglietta, M. M., D. Mastrangelo, and D. Conte, 2014: Influence of physics parameterization schemes on the simulation of a tropical-like cyclone in the Mediterranean Sea. *Atmos. Res.*

Miglietta, M. M., A. Moscatello, D. Conte, G. Mannarini, G. Lacorata, and R. Rotunno, 2011: Numerical analysis of a Mediterranean "hurricane" over south-eastern Italy: sensitivity experiments to sea surface temperature. *Atmos. Res.*, **101 (1)**, 412–426.

Millán, M., and Coauthors, 2005: Climatic feedbacks and desertification: the Mediterranean model. *J. Climate*, **18 (5)**, 684–701.

Molteni, F., R. Buizza, C. Marsigli, A. Montani, F. Nerozzi, and T. Paccagnella, 2001: A strategy for high-resolution ensemble prediction. I: Definition of representative members and global-model experiments. *Q. J. Roy. Meteor. Soc.*, **127 (576)**, 2069–2094.

Monahan, E. C., D. E. Spiel, and K. L. Davidson, 1986: A model of marine aerosol generation via whitecaps and wave disruption. *Oceanic whitecaps*, Springer, 167–174.

Montani, A., D. Cesari, C. Marsigli, and T. Paccagnella, 2011: Seven years of activity in the field of mesoscale ensemble forecasting by the COSMO-LEPS system: main achievements and open challenges. *Tellus A*, **63 (3)**, 605–624.

Montgomery, D. C., 2008: *Design and analysis of experiments*. 7th ed., John Wiley & Sons, United States of America, 680 p.

Morcrette, J.-J., 1991: Evaluation of model-generated cloudiness: Satellite-observed and model-generated diurnal variability of brightness temperature. *Mon. Weather Rev.*, **119 (5)**, 1205–1224.

Morrison, H., 2012: On the robustness of aerosol effects on an idealized supercell storm simulated with a cloud system-resolving model. *Atmos. Chem. Phys.*, **12 (16)**, 7689–7705.

Mühr, B., 2011: Wettergefahren Frühwarnung Website. http://www.wettergefahren-fruehwarnung.de/Artikel/20111103.html. Accessed: 2014-09-16.

Noppel, H., U. Blahak, A. Seifert, and K. D. Beheng, 2010: Simulations of a hailstorm and the impact of CCN using an advanced two-moment cloud microphysical scheme. *Atmos. Res.*, **96 (2)**, 286–301.

Palmen, E., 1948: On the formation and structure of tropical hurricanes. *Geophysica*, **3 (1)**, 26–38.

Parkinson, C. L., 2003: Aqua: An Earth-observing satellite mission to examine water and other climate variables. *Ieee. T. Geosci. Remote.*, **41 (2)**, 173–183.

Pettersen, S., 1956: *Weather Analysis and Forecasting*, Vol. 1. 2nd ed., McGraw-Hill, New York.

Phillips, V. T., T. W. Choularton, A. M. Blyth, and J. Latham, 2002: The influence of aerosol concentrations on the glaciation and precipitation of a cumulus cloud. *Q. J. Roy. Meteor. Soc.*, **128 (581)**, 951–971.

Phillips, V. T., P. J. DeMott, and C. Andronache, 2008: An empirical parameterization of heterogeneous ice nucleation for multiple chemical species of aerosol. *J. Atmos. Sci.*, **65 (9)**, 2757–2783.

Picornell, M., J. Campins, and A. Jansà, 2014: Detection and thermal description of medicanes from numerical simulation. *Nat. Hazard. Earth. Sys.*, **14 (5)**, 1059–1070.

Platnick, S., M. D. King, S. A. Ackerman, W. P. Menzel, B. A. Baum, J. C. Riédi, and R. A. Frey, 2003: The MODIS cloud products: Algorithms and examples from Terra. *Ieee. T. Geosci. Remote.*, **41 (2)**, 459–473.

Rasmussen, E., and C. Zick, 1987: A subsynoptic vortex over the Mediterranean with some resemblance to polar lows. *Tellus A*, **39 (4)**, 408–425.

Rieger, D., M. Bangert, C. Kottmeier, H. Vogel, and B. Vogel, 2014: Impact of aerosol on post-frontal convective clouds over Germany. *Tellus B*, **66**.

Riemer, N., H. Vogel, B. Vogel, and F. Fiedler, 2003: Modeling aerosols on the mesoscale-γ: Treatment of soot aerosol and its radiative effects. *J. Geophys. Res.-Atmos.*, **108 (D19)**.

Ritter, B., and J.-F. Geleyn, 1992: A comprehensive radiation scheme for numerical weather prediction models with potential applications in climate simulations. *Mon. Weather Rev.*, **120 (2)**, 303–325.

Rodrıguez, S., X. Querol, A. Alastuey, G. Kallos, and O. Kakaliagou, 2001: Saharan dust contributions to PM10 and TSP levels in Southern and Eastern Spain. *Atmos. Environ.*, **35 (14)**, 2433–2447.

Rodrıguez, S., X. Querol, A. Alastuey, and F. Plana, 2002: Sources and processes affecting levels and composition of atmospheric aerosol in the western Mediterranean. *J. Geophys. Res.-Atmos.*, **107 (D24)**, AAC 12–1–AAC 12–14.

Romero, R., and K. Emanuel, 2013: Medicane risk in a changing climate. *J. Geophys. Res.-Atmos.*, **118 (12)**, 5992–6001.

Rosenfeld, D., 1999: TRMM observed first direct evidence of smoke from forest fires inhibiting rainfall. *Geophys. Res. Lett.*, **26 (20)**, 3105–3108.

Rosenfeld, D., U. Lohmann, G. B. Raga, C. D. O'Dowd, M. Kulmala, S. Fuzzi, A. Reissell, and M. O. Andreae, 2008: Flood or drought: how do aerosols affect precipitation? *Science*, **321 (5894)**, 1309–1313.

Rosenfeld, D., W. L. Woodley, A. Khain, W. R. Cotton, G. Carrió, I. Ginis, and J. H. Golden, 2012: Aerosol effects on microstructure and intensity of tropical cyclones. *B. A.M. Meteorol. Soc.*, **93 (7)**, 987–1001.

Saunders, R., M. Matricardi, and P. Brunel, 1999: An improved fast radiative transfer model for assimilation of satellite radiance observations. *Q. J. Roy. Meteor. Soc.*, **125 (556)**, 1407–1425.

Schättler, U., G. Doms, and M. Baldauf, 2011: A Description of the Non-hydrostatic Regional COSMO-Model Part I: Dynamics and Numerics. Tech. rep., Deutscher Wetterdienst.

Schmetz, J., P. Pili, S. Tjemkes, D. Just, J. Kerkmann, S. Rota, and A. Ratier, 2002: An introduction to Meteosat second generation (MSG). *B. A.M. Meteorol. Soc.*, **83 (7)**, 977–992.

Seifert, A., and K. Beheng, 2006: A two-moment cloud microphysics parameterization for mixed-phase clouds. Part 1: Model description. *Meteorol. Atmos. Phys.*, **92 (1-2)**, 45–66.

Seifert, A., C. Köhler, and K. Beheng, 2012: Aerosol-cloud-precipitation effects over Germany as simulated by a convective-scale numerical weather prediction model. *Atmos. Chem. Phys.*, **12 (2)**, 709–725.

Seinfeld, J. H., and S. N. Pandis, 2012: *Atmospheric chemistry and physics: from air pollution to climate change.* John Wiley & Sons, United States of America, 1232 p.

Shao, Y., and H. Lu, 2000: A simple expression for wind erosion threshold friction velocity. *J. Geophys. Res.-Atmos.*, **105 (D17)**, 22 437–22 443.

Smith, M., P. Park, and I. Consterdine, 1993: Marine aerosol concentrations and estimated fluxes over the sea. *Q. J. Roy. Meteor. Soc.*, **119 (512)**, 809–824.

Song, C. H., and G. R. Carmichael, 1999: The aging process of naturally emitted aerosol (sea-salt and mineral aerosol) during long range transport. *Atmos. Environ.*, **33 (14)**, 2203–2218.

Squires, P., 1958: The microstructure and colloidal stability of warm clouds. *Tellus*, **10 (2)**, 262–271.

Stanelle, T., B. Vogel, H. Vogel, D. Bäumer, and C. Kottmeier, 2010: Feedback between dust particles and atmospheric processes over West Africa during dust episodes in March 2006 and June 2007. *Atmos. Chem. Phys.*, **10 (22)**, 10 771–10 788.

Stein, U., and P. Alpert, 1993: Factor separation in numerical simulations. *J. Atmos. Sci.*, **50 (14)**, 2107–2115.

Stevens, B., and O. Boucher, 2012: Climate science: The aerosol effect. *Nature*, **490 (7418)**, 40–41.

Stevens, B., and G. Feingold, 2009: Untangling aerosol effects on clouds and precipitation in a buffered system. *Nature*, **461 (7264)**, 607–613.

Stocker, T. F., and Coauthors, 2013: Climate change 2013: The physical science basis. *Intergovernmental Panel on Climate Change, Working Group I Contribution to the IPCC Fifth Assessment Report (AR5)(Cambridge Univ Press, New York).*

Stockwell, W. R., F. Kirchner, M. Kuhn, and S. Seefeld, 1997: A new mechanism for regional atmospheric chemistry modeling. *J. Geophys. Res.-Atmos.*, **102 (D22)**, 25 847–25 879.

Stockwell, W. R., P. Middleton, J. S. Chang, and X. Tang, 1990: The second generation regional acid deposition model chemical mechanism for regional air quality modeling. *J. Geophys. Res.-Atmos.*, **95 (D10)**, 16 343–16 367.

Tao, W.-K., J.-P. Chen, Z. Li, C. Wang, and C. Zhang, 2012: Impact of aerosols on convective clouds and precipitation. *Rev. Geophys.*, **50 (2)**.

Tao, W.-K., X. Li, A. Khain, T. Matsui, S. Lang, and J. Simpson, 2007: Role of atmospheric aerosol concentration on deep convective precipitation: Cloud-resolving model simulations. *J. Geophys. Res.-Atmos.*, **112 (D24)**.

Teller, A., and Z. Levin, 2006: The effects of aerosols on precipitation and dimensions of subtropical clouds: a sensitivity study using a numerical cloud model. *Atmos. Chem. Phys.*, **6 (1)**, 67–80.

Teller, A., and Z. Levin, 2008: Factorial method as a tool for estimating the relative contribution to precipitation of cloud microphysical processes and environmental conditions: Method and application. *J. Geophys. Res.-Atmos.*, **113 (D2)**.

Textor, C., and Coauthors, 2006: Analysis and quantification of the diversities of aerosol life cycles within AeroCom. *Atmos. Chem. Phys.*, **6 (7)**, 1777–1813.

The Meteo Company B.V., 2011: sat24.com Website. http://www.sat24.com. Accessed: 2014-09-16.

Tiedtke, M., 1989: A comprehensive mass flux scheme for cumulus parameterization in large-scale models. *Mon. Weather Rev.*, **117 (8)**, 1779–1800.

Tous, M., and R. Romero, 2013: Meteorological environments associated with medicane development. *Geosci. Model Dev.*, **33 (1)**, 1–14.

Tous, M., R. Romero, and C. Ramis, 2013: Surface heat fluxes influence on medicane trajectories and intensification. *Atmos. Res.*, **123**, 400–411.

Twomey, S., 1977: The influence of pollution on the shortwave albedo of clouds. *J. Atmos. Sci.*, **34 (7)**, 1149–1152.

Twomey, S. A., M. Piepgrass, and T. Wolfe, 1984: An assessment of the impact of pollution on global cloud albedo. *Tellus B*, **36 (5)**, 356–366.

University of Wisconsin-Madison, 2011: Space Science and Engineering Center. http://www.ssec.wisc.edu. Accessed: 2014-09-16.

van den Heever, S. C., G. G. Carrió, W. R. Cotton, P. J. DeMott, and A. J. Prenni, 2006: Impacts of nucleating aerosol on Florida storms. Part I: Mesoscale simulations. *J. Atmos. Sci.*, **63 (7)**, 1752–1775.

Van Den Heever, S. C., and W. R. Cotton, 2007: Urban aerosol impacts on downwind convective storms. *J. Appl. Meteorol. Clim.*, **46 (6)**, 828–850.

van den Heever, S. C., G. L. Stephens, and N. B. Wood, 2011: Aerosol indirect effects on tropical convection characteristics under conditions of radiative-convective equilibrium. *J. Atmos. Sci.*, **68 (4)**, 699–718.

Vogel, B., C. Hoose, H. Vogel, and C. Kottmeier, 2006: A model of dust transport applied to the Dead Sea area. *Meteorol. Z.*, **15 (6)**, 611–624.

Vogel, B., H. Vogel, D. Bäumer, M. Bangert, K. Lundgren, R. Rinke, and T. Stanelle, 2009: The comprehensive model system COSMO-ART - Radiative impact of aerosol on the state of the atmosphere on the regional scale. *Atmos. Chem. Phys.*, **9 (22)**, 8661–8680.

von Storch, H., H. Langenberg, and F. Feser, 2000: A spectral nudging technique for dynamical downscaling purposes. *Mon. Weather Rev.*, **128 (10)**, 3664–3673.

Walsh, K., F. Giorgi, and E. Coppola, 2014: Mediterranean warm-core cyclones in a warmer world. *Clim. Dynam.*, **42 (3-4)**, 1053–1066.

Wang, C., 2005: A modeling study of the response of tropical deep convection to the increase of cloud condensation nuclei concentration: 1. Dynamics and microphysics. *J. Geophys. Res.-Atmos.*, **110 (D21)**.

Wang, H., T. Auligné, and H. Morrison, 2012: Impact of microphysics scheme complexity on the propagation of initial perturbations. *Mon. Weather Rev.*, **140 (7)**, 2287–2296.

Wang, Y., K.-H. Lee, Y. Lin, M. Levy, and R. Zhang, 2014: Distinct effects of anthropogenic aerosols on tropical cyclones. *Nature Climate Change*.

White, B. R., 1979: Soil transport by winds on Mars. *J. Geophys. Res.-Sol. Ea.*, **84 (B9)**, 4643–4651.

Wilks, D., 1995: *Statistical Methods in the Atmospheric Sciences: An Introduction*. International geophysics series, Academic Press, United States of America, 704 p.

Zender, C. S., R. Miller, and I. Tegen, 2004: Quantifying mineral dust mass budgets: Terminology, constraints, and current estimates. *Eos, Transactions American Geophysical Union*, **85 (48)**, 509–512.

Zhang, H., G. M. McFarquhar, W. R. Cotton, and Y. Deng, 2009: Direct and indirect impacts of Saharan dust acting as cloud condensation nuclei on tropical cyclone eyewall development. *Geophys. Res. Lett.*, **36 (6)**.

Zhang, H.-M., J. J. Bates, and R. W. Reynolds, 2006: Assessment of composite global sampling: Sea surface wind speed. *Geophys. Res. Lett.*, **33 (17)**.

Zhang, Z., A. S. Ackerman, G. Feingold, S. Platnick, R. Pincus, and H. Xue, 2012: Effects of cloud horizontal inhomogeneity and drizzle on remote sensing of cloud droplet effective radius: Case studies based on large-eddy simulations. *J. Geophys. Res.-Atmos.*, **117 (D19)**.

Zinner, T., H. Mannstein, and A. Tafferner, 2008: Cb-TRAM: Tracking and monitoring severe convection from onset over rapid development to mature phase using multi-channel Meteosat-8 SEVIRI data. *Meteorol. Atmos. Phys.*, **101 (3-4)**, 191–210.

Acknowledgments

This doctoral research study was performed from 2013 to 2015 at the Institute for Meteorology and Climate Research, Troposphere Research (IMK-TRO) at the Karlsruhe Institute of Technology (KIT). It took place in the working group "Aerosols, Trace Gases and Climate Processes" under the leadership of Dr. Bernhard Vogel. I would like to thank everybody who supported me during this study.

I am very grateful to my supervisor Prof. Christoph Kottmeier for his support during my entire time at the KIT. By his confidence, I was able to join many conferences and present my work there. I also thank Prof. Corinna Hoose for accepting the co-supervision of this work and giving me valuable tips.

I would like to express my deep gratitude to Dr. Bernhard Vogel, who supported me in his working group for so many years. He always encouraged me scientifically and had an ear for my questions and concerns. The discussions with him were most fruitful and gave clear ways how to proceed.

Also, I would like to cordially thank Dr. Heike Vogel for taking much of her time to discuss and solve problems and Daniel Rieger for reading and reviewing most of this thesis. It's great to have a friend sitting next to myself in the office. In addition, I want to acknowledge the rest of the working group, Tobias Schad, Carolin Walter, Rowell Hagemann, and Konrad Deetz as well as the former members Dr. Max Bangert, Dr. Kristina Lundgren, and Dr. Andrew Ferrone. You always listened to current trouble concerning my research and proposed new ways of looking at problems.

I want to express my gratitude to Dr. Christian Keil from the Ludwig-Maximilians-Universität in Munich, who gave me the opportunity to use the Displacement and Amplitude Score. I am also thankful to PhD Chiara

Marsigli and PhD Andrea Montani from the Regional Agency for Environmental Protection in the Emilia-Romagna region, Italy, who kindly provided me with the COSMO-LEPS data.

Moreover, I am very grateful to the non-scientific staff, especially to Gabi Klinck, Rosalba Gräbner, Silvia Deckert, and Doris Stenschke. They set the administrative and technical baselines to create an environment making research at the institute both possible and fun.

Besides, I would like to acknowledge the Helmholtz Climate Initiative REKLIM, which funded my work over the last three years and the Graduate school for Climate and Environment at the KIT, GRACE, which gave me the possibility to enrich my knowledge by distinct courses and supported my participation in several conferences.

Last but not least, I am deeply grateful to my family, friends and my boyfriend. You did not only stand behind me and were sympathetic but also gave me a lot of strength. Without you this thesis wouldn't exist.

Wissenschaftliche Berichte des Instituts für Meteorologie und Klimaforschung des Karlsruher Instituts für Technologie (0179-5619)

Bisher erschienen:

Nr. 1: *Fiedler, F. / Prenosil, T.*
Das MESOKLIP-Experiment. (Mesoskaliges Klimaprogramm im Oberrheintal). August 1980

Nr. 2: *Tangermann-Dlugi, G.*
Numerische Simulationen atmosphärischer Grenzschichtströmungen über langgestreckten mesoskaligen Hügelketten bei neutraler thermischer Schichtung. August 1982

Nr. 3: *Witte, N.*
Ein numerisches Modell des Wärmehaushalts fließender Gewässer unter Berücksichtigung thermischer Eingriffe. Dezember 1982

Nr. 4: *Fiedler, F. / Höschele, K. (Hrsg.)*
Prof. Dr. Max Diem zum 70. Geburtstag. Februar 1983 (vergriffen)

Nr. 5: *Adrian, G.*
Ein Initialisierungsverfahren für numerische mesoskalige Strömungsmodelle. Juli 1985

Nr. 6: *Dorwarth, G.*
Numerische Berechnung des Druckwiderstandes typischer Geländeformen. Januar 1986

Nr. 7: *Vogel, B.; Adrian, G. / Fiedler, F.*
MESOKLIP-Analysen der meteorologischen Beobachtungen von mesoskaligen Phänomenen im Oberrheingraben. November 1987

Nr. 8: *Hugelmann, C.-P.*
Differenzenverfahren zur Behandlung der Advektion. Februar 1988

Nr. 9: *Hafner, T.*
Experimentelle Untersuchung zum Druckwiderstand der Alpen. April 1988

Nr. 10: *Corsmeier, U.*
Analyse turbulenter Bewegungsvorgänge in der maritimen atmosphärischen Grenzschicht. Mai 1988

Nr. 11: *Walk, O. / Wieringa, J.(eds)*
Tsumeb Studies of the Tropical Boundary-Layer Climate. Juli 1988

Nr. 12: *Degrazia, G. A.*
Anwendung von Ähnlichkeitsverfahren auf die turbulente Diffusion in der konvektiven und stabilen Grenzschicht. Januar 1989

Nr. 13: *Schädler, G.*
Numerische Simulationen zur Wechselwirkung zwischen Landoberflächen und atmophärischer Grenzschicht. November 1990

Nr. 14: *Heldt, K.*
Untersuchungen zur Überströmung eines mikroskaligen Hindernisses in der Atmosphäre. Juli 1991

Nr. 15: *Vogel, H.*
Verteilungen reaktiver Luftbeimengungen im Lee einer Stadt – Numerische Untersuchungen der relevanten Prozesse. Juli 1991

Nr. 16: *Höschele, K.(ed.)*
Planning Applications of Urban and Building Climatology – Proceedings of the IFHP / CIB-Symposium Berlin, October 14-15, 1991. März 1992

Nr. 17: *Frank, H. P.*
Grenzschichtstruktur in Fronten. März 1992

Nr. 18: *Müller, A.*
Parallelisierung numerischer Verfahren zur Beschreibung von Ausbreitungs- und chemischen Umwandlungsprozessen in der atmosphärischen Grenzschicht. Februar 1996

Nr. 19: *Lenz, C.-J.*
Energieumsetzungen an der Erdoberfläche in gegliedertem Gelände. Juni 1996

Nr. 20: *Schwartz, A.*
Numerische Simulationen zur Massenbilanz chemisch reaktiver Substanzen im mesoskaligen Bereich. November 1996

Nr. 21: *Beheng, K. D.*
Professor Dr. Franz Fiedler zum 60. Geburtstag. Januar 1998

Nr. 22: *Niemann, V.*
Numerische Simulation turbulenter Scherströmungen mit einem Kaskadenmodell. April 1998

Nr. 23: *Koßmann, M.*
Einfluß orographisch induzierter Transportprozesse auf die Struktur der atmosphärischen Grenzschicht und die Verteilung von Spurengasen. April 1998

Nr. 24: *Baldauf, M.*
Die effektive Rauhigkeit über komplexem Gelände – Ein Störungstheoretischer Ansatz. Juni 1998

Nr. 25: *Noppel, H.*
Untersuchung des vertikalen Wärmetransports durch die Hangwindzirkulation auf regionaler Skala. Dezember 1999

Nr. 26: *Kuntze, K.*
Vertikaler Austausch und chemische Umwandlung von Spurenstoffen über topographisch gegliedertem Gelände. Oktober 2001

Nr. 27: *Wilms-Grabe, W.*
Vierdimensionale Datenassimilation als Methode zur Kopplung zweier verschiedenskaliger meteorologischer Modellsysteme. Oktober 2001

Nr. 28: *Grabe, F.*
Simulation der Wechselwirkung zwischen Atmosphäre, Vegetation und Erdoberfläche bei Verwendung unterschiedlicher Parametrisierungsansätze. Januar 2002

Nr. 29: *Riemer, N.*
Numerische Simulationen zur Wirkung des Aerosols auf die troposphärische Chemie und die Sichtweite. Mai 2002

Nr. 30: *Braun, F. J.*
Mesoskalige Modellierung der Bodenhydrologie. Dezember 2002

Nr. 31: *Kunz, M.*
Simulation von Starkniederschlägen mit langer Andauer über Mittelgebirgen. März 2003

Nr. 32: *Bäumer, D.*
Transport und chemische Umwandlung von Luftschadstoffen im Nahbereich von Autobahnen – numerische Simulationen. Juni 2003

Nr. 33: *Barthlott, C.*
Kohärente Wirbelstrukturen in der atmosphärischen Grenzschicht. Juni 2003

Nr. 34: *Wieser, A.*
Messung turbulenter Spurengasflüsse vom Flugzeug aus. Januar 2005

Nr. 35: *Blahak, U.*
Analyse des Extinktionseffektes bei Niederschlagsmessungen mit einem C-Band Radar anhand von Simulation und Messung. Februar 2005

Nr. 36: *Bertram, I.*
Bestimmung der Wasser- und Eismasse hochreichender konvektiver Wolken anhand von Radardaten, Modellergebnissen und konzeptioneller Betrachtungen. Mai 2005

Nr. 37: *Schmoeckel, J.*
Orographischer Einfluss auf die Strömung abgeleitet aus Sturmschäden im Schwarzwald während des Orkans „Lothar". Mai 2006

Nr. 38: *Schmitt, C.*
Interannual Variability in Antarctic Sea Ice Motion: Interannuelle Variabilität antarktischer Meereis-Drift. Mai 2006

Nr. 39: *Hasel, M.*
Strukturmerkmale und Modelldarstellung der Konvektion über Mittelgebirgen. Juli 2006

Ab Band 40 erscheinen die Wissenschaftlichen Berichte des Instituts für Meteorologie und Klimaforschung bei KIT Scientific Publishing (ISSN 0179-5619). Die Bände sind unter www.ksp.kit.edu als PDF frei verfügbar oder als Druckausgabe bestellbar.

Nr. 40: *Lux, R.*
Modellsimulationen zur Strömungsverstärkung von orographischen Grundstrukturen bei Sturmsituationen
ISBN 978-3-86644-140-8

Nr. 41: *Straub, W.*
Der Einfluss von Gebirgswellen auf die Initiierung und Entwicklung konvektiver Wolken
ISBN 978-3-86644-226-9

Nr. 42: *Meißner, C.*
High-resolution sensitivity studies with the regional climate model COSMO-CLM
ISBN 978-3-86644-228-3

Nr. 43: *Höpfner, M.*
Charakterisierung polarer stratosphärischer Wolken mittels hochauflösender Infrarotspektroskopie
ISBN 978-3-86644-294-8

Nr. 44: *Rings, J.*
Monitoring the water content evolution of dikes
ISBN 978-3-86644-321-1

Nr. 45: *Riemer, M.*
Außertropische Umwandlung tropischer Wirbelstürme : Einfluss auf das Strömungsmuster in den mittleren Breiten
ISBN 978-3-86644-766-0

Nr. 46: *Anwender, D.*
Extratropical Transition in the Ensemble Prediction System of the ECMWF : Case Studies and Experiments
ISBN 978-3-86644-767-7

Nr. 47: *Rinke, R.*
Parametrisierung des Auswaschens von Aerosolpartikeln durch Niederschlag
ISBN 978-3-86644-768-4

Nr. 48: *Stanelle, T.*
Wechselwirkungen von Mineralstaubpartikeln mit thermodynamischen und dynamischen Prozessen in der Atmosphäre über Westafrika
ISBN 978-3-86644-769-1

Nr. 49: *Peters, T.*
Ableitung einer Beziehung zwischen der Radarreflektivität, der Niederschlagsrate und weiteren aus Radardaten abgeleiteten Parametern unter Verwendung von Methoden der multivariaten Statistik
ISBN 978-3-86644-323-5

Nr. 50: *Khodayar Pardo, S.*
High-resolution analysis of the initiation of deep convection forced by boundary-layer processes
ISBN 978-3-86644-770-7

Nr. 51: *Träumner, K.*
Einmischprozesse am Oberrand der konvektiven atmosphärischen Grenzschicht
ISBN 978-3-86644-771-4

Nr. 52: *Schwendike, J.*
Convection in an African Easterly Wave over West Africa and the Eastern Atlantic : A Model Case Study of Hurricane Helene (2006) and its Interaction with the Saharan Air Layer
ISBN 978-3-86644-772-1

Nr. 53: *Lundgren, K.*
Direct Radiative Effects of Sea Salt on the Regional Scale
ISBN 978-3-86644-773-8

Nr. 54: *Sasse, R.*
Analyse des regionalen atmosphärischen Wasserhaushalts unter Verwendung von COSMO-Simulationen und GPS-Beobachtungen
ISBN 978-3-86644-774-5

Nr. 55: *Grenzhäuser, J.*
Entwicklung neuartiger Mess- und Auswertungsstrategien für ein scannendes Wolkenradar und deren Anwendungsbereiche
ISBN 978-3-86644-775-2

Nr. 56: *Grams, C.*
Quantification of the downstream impact of extratropical transition for Typhoon Jangmi and other case studies
ISBN 978-3-86644-776-9

Nr. 57: *Keller, J.*
Diagnosing the Downstream Impact of Extratropical Transition Using Multimodel Operational Ensemble Prediction Systems
ISBN 978-3-86644-984-8

Nr. 58: *Mohr, S.*
Änderung des Gewitter- und Hagelpotentials im Klimawandel
ISBN 978-3-86644-994-7

Nr. 59: *Puskeiler, M.*
Radarbasierte Analyse der Hagelgefährdung in Deutschland
ISBN 978-3-7315-0028-5

Nr. 60: *Zeng, Y.*
Efficient Radar Forward Operator for Operational Data Assimilation within the COSMO-model
ISBN 978-3-7315-0128-2

Nr. 61: *Bangert, M. J.*
Interaction of Aerosol, Clouds, and Radiation on the Regional Scale
ISBN 978-3-7315-0123-7

Nr. 62: *Jerger, D.*
Radar Forward Operator for Verification of Cloud Resolving Simulations within the COSMO Model
ISBN 978-3-7315-0172-5

Nr. 63: *Maurer, V.*
Vorhersagbarkeit konvektiver Niederschläge : Hochauflösende Ensemblesimulationen für Westafrika
ISBN 978-3-7315-0189-3

Nr. 64: *Stawiarski, C.*
Optimizing Dual-Doppler Lidar Measurements of Surface Layer Coherent Structures with Large-Eddy Simulations
ISBN 978-3-7315-0197-8

Nr. 65: *Mahlke, H.*
Mechanismen der Auslösung hochreichender Konvektion im südwestdeutschen Mittelgebirgsraum
ISBN 978-3-7315-0203-6

Nr. 66: *Fosser, G.*
Precipitation statistics from regional climate model at resolutions relevant for soil erosion
ISBN 978-3-7315-0227-2

Nr. 67: *Adler, B.*
Boundary-Layer Processes Producing Mesoscale Water-Vapour Variability over a Mountainous Island
ISBN 978-3-7315-0247-0

Nr. 68: *Kraut, I.*
Separating the Aerosol Effect in Case of a „Medicane"
ISBN 978-3-7315-0405-4